Penguins

Penguins

The Ultimate Guide

Second Edition

Tui De Roy • Mark Jones • Julie Cornthwaite

PRINCETON UNIVERSITY PRESS

Princeton and Oxford

Published in 2022 by Princeton University Press
41 William Street, Princeton, New Jersey 08540
99 Banbury Road, Oxford OX2 6JX
press.princeton.edu

First published by David Bateman Ltd, 2013
2/5 Workspace Drive, Hobsonville, Auckland, New Zealand
www.batemanbooks.co.nz

Library of Congress Control Number 2021947827

ISBN 978-0-691-23357-4
ebook ISBN 978-0-691-23358-1

Project editor: Tracey Borgfeldt
Design: Trevor Newman / Maps: Tim Nolan Black Ant

British Library Cataloging-in-Publication Data is available

Printed in China by Everbest Printing Co. Ltd

10 9 8 7 6 5 4 3 2 1

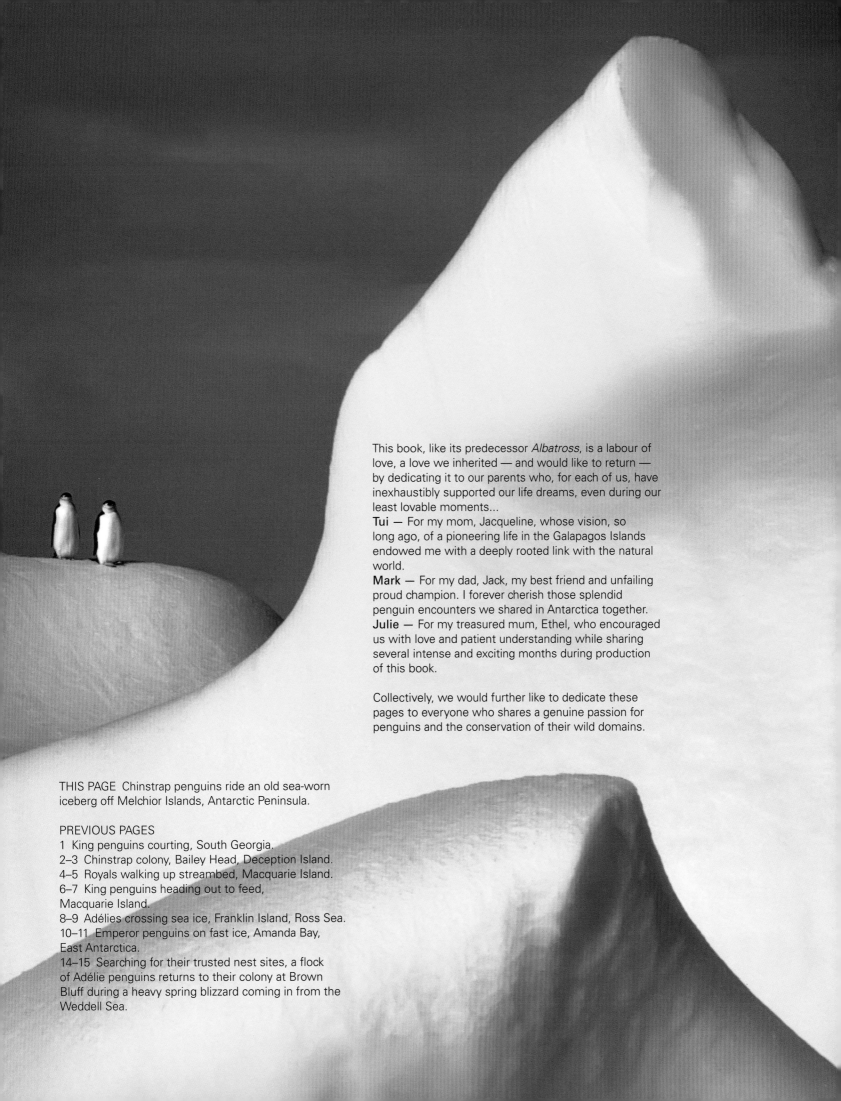

This book, like its predecessor *Albatross*, is a labour of love, a love we inherited — and would like to return — by dedicating it to our parents who, for each of us, have inexhaustibly supported our life dreams, even during our least lovable moments...

Tui — For my mom, Jacqueline, whose vision, so long ago, of a pioneering life in the Galapagos Islands endowed me with a deeply rooted link with the natural world.

Mark — For my dad, Jack, my best friend and unfailing proud champion. I forever cherish those splendid penguin encounters we shared in Antarctica together.

Julie — For my treasured mum, Ethel, who encouraged us with love and patient understanding while sharing several intense and exciting months during production of this book.

Collectively, we would further like to dedicate these pages to everyone who shares a genuine passion for penguins and the conservation of their wild domains.

THIS PAGE Chinstrap penguins ride an old sea-worn iceberg off Melchior Islands, Antarctic Peninsula.

PREVIOUS PAGES
1 King penguins courting, South Georgia.
2–3 Chinstrap colony, Bailey Head, Deception Island.
4–5 Royals walking up streambed, Macquarie Island.
6–7 King penguins heading out to feed,
Macquarie Island.
8–9 Adélies crossing sea ice, Franklin Island, Ross Sea.
10–11 Emperor penguins on fast ice, Amanda Bay,
East Antarctica.
14–15 Searching for their trusted nest sites, a flock of Adélie penguins returns to their colony at Brown Bluff during a heavy spring blizzard coming in from the Weddell Sea.

CONTENTS

1   LIFE BETWEEN TWO WORLDS   20
Tui De Roy

2   SCIENCE AND CONSERVATION   146
Mark Jones

3 SPECIES NATURAL HISTORY   190
Julie Cornthwaite

ABOVE An Adélie penguin hurries back to its nest through an Antarctic blizzard, Weddell Sea.
RIGHT Courting King penguins, Falkland Islands.
BELOW Mark is besieged by curious Royal penguins, Macquarie Island.

'Personally, I find penguins a tremendous source of inspiration. Not only are they superb examples of some of the animal kingdom's most extraordinary adaptations, but watching them tackle the daunting challenges of their everyday life with exuberant gusto makes me feel that nothing is unattainable as long as you throw yourself into it heart and soul. Mark, Julie and I, plus our many specialist contributors, have all experienced innumerable poignant and intense moments with penguins, providing us with profound insights into their world and their ways. We hope you will find them every bit as fascinating and enjoyable as we do.'

Tui De Roy

Prologue: Penguin Passion

People have been fascinated by penguins for just about as long as we've known they existed. For most of us, simply watching penguins elicits smiles, giggles, praise and sympathy, perhaps because their antics so often remind us of our own. When penguins are on land, their actions appear to us so humorous and expressive that we can be excused for thinking we understand them perfectly, and therefore identify with what looks like moods and foibles similar to our own. We forget that their private life is as complex and mysterious as that of any wild animal or that the bulk of their existence takes place very far from our prying eyes, hidden beneath the ocean waves.

Another popular misconception is that all penguins live around the poles. In reality, penguins are constrained to the southern hemisphere, but only four species form colonies along parts of the Antarctic coastline, remaining at least 1200 km (745 miles) from the South Pole. The entire 'crested dynasty' (seven species), live in slightly milder climates, mostly north of the Polar Front, nesting on subantarctic islands. More northerly still are the 'jackass'-type penguins (so-called because of their donkey-like braying) of South Africa and South America. Amongst their ranks is the only truly tropical species, the Galapagos penguin. This one even ventures into the northern hemisphere, crossing the equator. The diminutive Little blue, or Fairy, penguin, and the enigmatic Yellow-eyed penguin, make their homes in temperate regions of Australia and New Zealand. Thus we can find desert penguins, forest penguins and even night penguins; solitary penguins and gregarious penguins; 30 kg (66 lb) giants and 1 kg (2.2 lb) midgets.

While a few types of penguins are relatively well-known — thanks to zoos, books, films and select travel destinations — not everyone realises the family is represented by 18 species (or even 19, depending on the taxonomy used). Over half of these live in places so remote that they are rarely observed or photographed, and therefore poorly documented.

So it is that when my co-authors and I set out to produce this book, we first wanted to ensure that we would have ample time to become intimately acquainted with every one of the world's penguin species, no matter how difficult to find. Spanning over 15 years, between us, we've logged a veritable cornucopia of memorable worldwide penguin encounters. Here are just a few examples.

Growing up in the Galapagos Islands, perhaps I enjoyed an unfair advantage. Galapagos penguins were my childhood neighbours, whose underwater realm I was able to share from time to time. Years later, having taken up residence in New Zealand, I once again found myself in the proximity of some lesser-known species. Here I have crawled through fern-padded forest undergrowth to spy on skittish Fiordland penguin family life, or hidden between megaherbs on Campbell Island to watch timid Yellow-eyed penguins trotting through fields of bright blossoms. On a damp Tasmanian beach, I sat up all night witnessing the surreptitious comings and goings of Little Blue penguins under the cover of darkness.

Some of Mark's fondest memories come from observing the great rush-hour traffic of Royal penguins on Australia's subantarctic Macquarie Island as they ducked between giant cabbage leaves while still finding time to detour for a close inspection of this strange alien, or watching cheeky Gentoos surreptitiously stealing stones from one another's nests. In contrast, Julie's favourite recollections are of curious King penguins making her feel like the Pied Piper on South Georgia Island, trailing an inquisitive retinue of large fluffy chicks down a beach, or strolling amid the vast Magellanic 'penguinopolis' of Punta Tombo in Argentina.

In the Falkland Islands, our attention was riveted by the dogged perseverance of Southern rockhoppers being tumbled and tossed by gigantic storm waves as they attempted landing at the foot of a 30-m (100-ft) cliff, which they then climbed using beaks and claws to reach their nesting colony far above.

In the frozen wastes of East Antarctica, I once walked in hushed silence alongside Emperor penguins tobogganing across the frozen sea, their long lines vanishing in the far distance between cathedral-like entrapped icebergs strafed by rays of fiery midnight sun. On tiny Tilgo Island along Chile's desert coast, I camped among Humboldt penguins nesting under tall cacti. I've marvelled at the ingenuity of African penguins establishing a successful new colony near Cape Town, using a coastal housing development as a shield against wild land predators. In the South Orkney Islands I shivered alongside huddled Adélie chicks, their light fluffy down and my thick parka equally sodden, ironically chilled when unseasonably warm temperatures turned snowfall to rain. And on faraway Gough Island, I laughed at the comical expressions of the Northern Rockhoppers, their flowing golden head plumes blowing about in the wind like mops.

Penguin watching offers infinitely more than reflections of ourselves. 🐧

ABOVE A busy Chinstrap penguin carries a small stone to its nest, important to ensure drainage around the eggs, Elephant Island.
BELOW Tui shares a private underwater moment with a Galapagos penguin, Bartolomé Island.
BELOW LEFT Julie followed by an excited gang of inquisitive King penguin chicks, South Georgia.

Global Distribution of Penguins

The world's 18 species of penguins are entirely confined to the southern hemisphere, yet only four of them actually nest on or near the Antarctic continent, while just one reaches the equator. Oceanographic features, such as summer and winter sea-ice cover and the position of the Antarctic Convergence (also known as the Polar Front), define the feeding and breeding ranges of many other penguin species. The 60th parallel represents the legal limit of Antarctica and the Southern Ocean, where human activities are governed by treaty. Place names on this map indicate main breeding sites, whereas red arrows point to photographic locations represented in this book, encompassing all penguin-rich regions over a period spanning more than two decades.

Tristan da Cunha Is

Goug

South Georgia

South Sandwich Is

ARGENTINA

CHILE

PERU

Atacama Desert islands

Sechura Desert islands

Patagonia

Falkland Is

South Orkney Is

Riiser-Larsen Ice Shelf

Anta

Elephant I
King George I

South Shetland Is

Antarctic Peninsula

Weddell Sea

Chiloé I
Chilean Fjords
Magellan Straits

Diego Ramirez Is

Petermann

Bellingshausen Sea

Peter 1 I

West Antarct

90°W
Galapagos Is

Amundsen Sea

Shepard I

Ross

Cape

SOUTHERN OCEAN

Sea-ice extent — Winter
Sea-ice extent — Summer
Antarctic Convergence
Locations photographed during the production of this book

Antipod
Boun

Chatham Is

N
ZEA

NAMIBIA

SOUTH
AFRICA

30°S

Prince Edward
& Marion Is

...vet I

Crozet Is

...UTHERN OCEAN

60°S

Lazarev Ice
Shelf

Kloa Point

Kerguelen Is

Heard &
McDonald Is

Amsterdam &
St Paul Is

Cape Darnley

Prydz Bay

East Antarctica

90°E

Windmill Is

Balleny Is

Macquarie I

Is

Auckland Is

Tasmania

Snares Is

Phillip I

AUSTRALIA

PENGUIN BREEDING
DISTRIBUTION
Many of the southern penguin
species have circumpolar
distributions, their respective
breeding ranges forming concentric
rings defined by preferred habitat.
For example, Emperors (OPPOSITE)
hug the continental ice girdle,
Adélies (ABOVE) stay close to
ice but use rocky ground between
54°S and 77°S, Kings (OPPOSITE
BELOW) choose islands along the
Antarctic Convergence, and Southern
rockhoppers (LEFT) stay mostly
north of it. Gentoos (BELOW) use
a much wider climatic range, from
the snowy Antarctic Peninsula to
subantarctic Falkland Islands. For
details, see individual species maps
in Part 3.

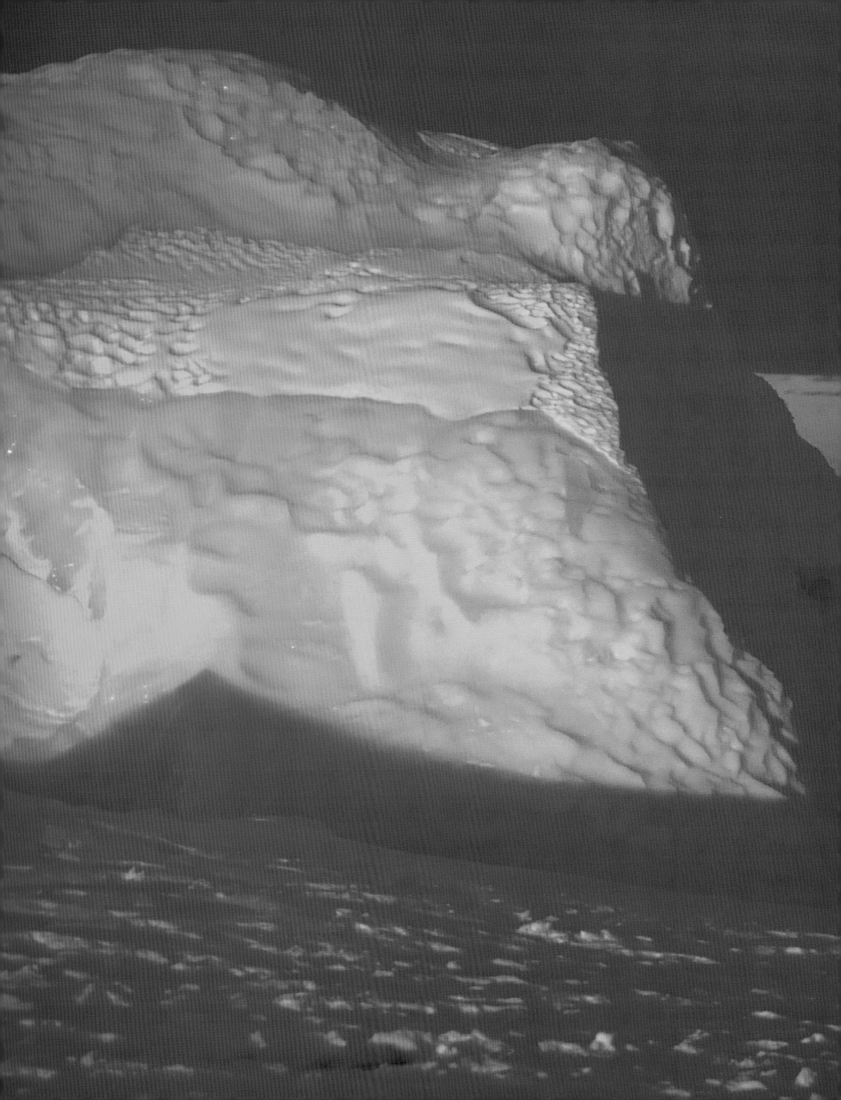

1
LIFE BETWEEN TWO WORLDS

Tui De Roy

Penguin Seasons: The Cycle of Life

It is early March deep in the Ross Sea of Antarctica, the end of the brief southern summer. The sky hangs low, heavy with inky-purple clouds, snow flurries skirmishing on the horizon. A fresh dusting already covers the still-sodden ground, and a rampart of jagged ice slabs — the remnants of once-smooth sea ice now driven ashore by angry waves — thumps heavily upon the black volcanic beach, grinding and creaking. Throughout this otherworldly scene thousands of little black-and-white figures run about, stumbling and scuttling, hopping and screeching, all heading toward the open water beyond the ice. Tens of thousands more surge forth from the hinterlands, flowing like rivers across the flats, and more still tumble down the steep icy slopes farther back.

Winter exodus

Cape Adare is home to the world's largest nesting colony of Adélie penguins, the southernmost of all penguin species. In a good year, a quarter of a million young penguins leave these shores, anxious to start a

new life at sea, a life of which they know nothing but the general notions their instincts dictate. Wide-eyed, most of them still sporting topknots and ragged tufts of fluffy down from their short babyhood, they are rushing into the unknown, about to undergo a change in lifestyle almost as drastic and wondrous as a caterpillar's metamorphosis into a butterfly, or a dragonfly nymph suddenly rising into the wind from swampy reed beds. The chicks' down, an efficient insulation only when dry, is just being replaced by new waterproof coats of feathers designed to keep them warm and dry underwater. The young penguins are only about 80% of their full-grown size and weight, but they're ready.

I sit and watch quietly for a while. More throngs gather, urged onward by the shrill calls of departing adults already beyond the ice barrier, porpoising and splashing through leads of dark water. The chicks clamber awkwardly onto rocking grounded bergy-bits, slipping and falling. Despite these difficulties, their eagerness to start travelling remains undampened,

ABOVE Swapping incubation duties, a King penguin pair gingerly rolls their precious egg from one set of feet to the other, Falkland Islands.
LEFT Porpoising at full speed, homeward bound Gentoo penguins swim through clear Falkland Islands waters.

PRECEDING PAGES Under the midnight sun, a lone Emperor penguin marches past entrapped icebergs over the frozen sea to reach the Kloa Point colony in East Antarctica.

as if their life depended on it. Which of course, it does. Their boundless excitement, their dogged determination, mounting as their numbers swell by the shore, becomes contagious. I feel that if they can succeed in making it through this coming of age, then no challenge would ever be too great to undertake.

Of course, many won't make it; that is the story of life on earth. Already a few dozen bedraggled stragglers — hatched just a little too late, or whose parents didn't find them quite as much food as their more experienced neighbours — sit back bewildered on their pebble nests, bereft of protective parents and neighbours. Lacking the plumage needed to take to sea, they are doomed as predatory skuas and giant petrels move in.

For those who are departing, the first hurdle comes at the overhanging ice edge, where they now catch their first glimpse of seawater. Utterly inexperienced but gung-ho nonetheless, they launch themselves, some headfirst, some feet first, but most just plain helter-skelter. Regardless, when they hit the water — oh surprise — they actually don't know how to swim. Wave after wave of young penguins plop into the sea, only to founder and flail, little flippers furiously flapping as they try to heft their bodies onto the surface of the water, rather than diving through it. Only when, after long minutes of no progress, they accidentally submerge do they suddenly discover what those fantastically adapted 'wings' were made for: underwater flight. From now on, they will only emerge briefly to breathe, and to preen and to replenish the air cushion trapped within their dense feathers.

All around Antarctica, similar scenes are taking place. The great Adélie penguin exodus is under way everywhere. Like human refugees fleeing from natural calamity, they must head north, staying slightly ahead of the big winter freeze. Chinstrap penguins are doing likewise along the Antarctic Peninsula and

adjacent islands, several degrees of latitude short of the polar circle and many thousands of kilometres from here. They will move to the open pack ice edge, their preferred winter habitat. And all around the girdle of subantarctic islands still farther north, millions of crested penguins of different species will shortly also be departing.

Amazingly, two species of penguins will move in the opposite direction from all others, heading south into winter darkness. King penguins nest along the periphery of Antarctica on islands strung close to the rich oceanic mixing zone called the Antarctic Convergence, or Polar Front. They breed year-round regardless of season because their large, slow-growing chicks need almost a year to fledge, but in winter the parents leave them unattended for several months, moving down to the edge of the pack-ice to feed. Here they may well rub shoulders with their giant cousin, the Emperor penguin, whose life habits in turn reach even greater extremes. As we will see in greater detail later in this book, the Emperor defies all rules by heading south into the depth of the Antarctic winter to lay its eggs and raise chicks perched on the surface of the frozen sea.

What makes a penguin?

Regardless of range and nesting habits, all penguins must undertake that great leap of faith, that biological dichotomy that is the difference between life at sea, for which they are superbly suited, and breeding on land, a task they tackle with remarkable ability and zest, considering that their body plan has not prioritised adaptations to function well in this environment.

Ashore, penguins remind us of ourselves in an amazing variety of ways: stance, gait, attire, personality, sociability, curiosity, hyperactivity and, of course, short tempers. But the projection of our own attitudes, which has endeared penguins to most

ABOVE A pair of courting Emperor penguins perform a 'slow bow' display, Weddell Sea.
BELOW Northern rockhoppers dry off after coming ashore, Nightingale Island, Tristan da Cunha.
BELOW RIGHT A male Adélie penguin advertises by 'sky-pointing' in the middle of the colony. Others around him are incubating, hunkered down against a blizzard, Weddell Sea.

humans and indeed found them a special place in our hearts, should never overshadow the wondrous animals they are in their own right: Extraordinary beings that abandoned the traditional lifestyle of birds not long after these had begun taking to the air, and headed back into the life-giving sea somewhere around the time of the dinosaurs' abrupt demise.

At sea, a penguin's greatest metabolic preoccupation is to keep warm, but on land very often the opposite applies. To achieve both, it is equipped with a remarkable array of adaptations. Densely packed, impermeable feathers that fit together like scales — the densest plumage of any bird — trap air in a plush layer of down undercoat, much like a diver's neoprene dry suit worn over fleece undergarments. This insulation is further backed up by a generous layer of fat beneath the skin acting as an additional thermal barrier. Yet on land, where overheating is a real threat, the feathers may be raised to allow free airflow between them, while panting for evaporative cooling

from the throat lining partially makes up for that blubbery subcutaneous envelope.

Penguin thermal controls do not stop there. A complex counter-current heat exchange circulatory system, including blood vessels that run in grooves along the wing bones, allows warmth to be shunted back to maintain a steady body core temperature while flippers and feet can be nearly as cold as the surrounding water. But on land the mechanism can be inverted, so that hot blood flushes to these scantily feathered extremities, turning them bright pink during hot weather or heavy exertion as excess heat is vented away.

Even penguin locomotion is twofold and fully amphibious. Those perfect appendages, super fit for rapid and agile underwater travel — flippers for propulsion and feet for steering — can quickly swap roles upon making landfall. Now wings serve for balance only, whereas those stubby feet become sturdy boots, armed with sharp crampons for ice travel

ABOVE King penguins stand tall during a group courtship session on the green spring grass at Volunteer Beach, Falkland Islands.

ABOVE As dusk falls, a
slow exposure outlines
the movements of
breaking surf and
beach-landing Southern
rockhopper penguins,
Saunders Island,
Falkland Islands.

and, in some species enable a springy pogo-stick-like bounce for jumping several body-lengths amongst large boulders.

Life at sea

Upon leaving the nesting grounds, penguins come into their own. Like anyone freed of family obligations, and no longer constrained by shuttlecock commuting patterns, many of them range over incredible distances. As they often like to remain sociable at sea, they use especially loud contact calls — harsh and piercing squawks never heard on land — that are clearly audible over wind and waves. Some species, in particular Adélies, Chinstraps and, to a lesser degree, Emperors, will travel with the drifting pack-ice, using floes to rest upon when not fishing. Yet Rockhoppers, Macaronis, Kings and many others head out into the open ocean, where they may range hundreds if not thousands of kilometres offshore. Porpoising like dolphins

when moving quickly, they will not return to land for many months. More than once I have been utterly dumbfounded when sailing through angry seas in the far south, to suddenly catch the unmistakable 'Kraaark' of a penguin call carrying over the roar of the wind, followed by a glimpse of little bullet-shaped bodies erupting briefly between frothy wave-crests.

The fastest swimming speed recorded is for the Gentoo penguin, clocked at 36 kph (22 mph), although it is likely that the sleek King could outpace it. The depth record goes to an Emperor penguin who logged an incredible 564 m (1850 ft) during a nine-minute dive, though most dives average a mere 100–120 m (330–395 ft) and last between three and six minutes. The longest dive duration also went to Emperor penguins, twice clocking a maximum of 22 minutes without a breath.

There is only one other requirement besides nesting that must bring penguins back to dry land: the moult. Unlike many other birds that shed a few feathers at

ABOVE RIGHT A Magellanic penguin paddles up a shallow stream before hopping ashore to reach the busy nesting colony at Seno Otway in Chilean Patagonia.
ABOVE LEFT Galapagos penguins head out in small groups on their daily fishing trip along the rocky lava shores of Bartolomé Island, Galapagos.
LEFT Porpoising enables travelling Gentoos to rise for a breath without losing speed as they head back to their chicks after a day's fishing, Antarctic Peninsula.

a time until their entire plumage has been replaced, penguins undergo what is termed a 'catastrophic moult'. In their world, it's all-or-nothing, since a partially clad penguin has no chance of staying warm while submerged. So they must come ashore, generally just before the onset of winter, having fattened up to the maximum to endure the long fast while their new feathers grow. Cold in their ragged coats, scruffy and hungry, they endure the indignities of being land-bound for about three to four weeks. Once the moult is completed, they will be as good as new for another year of ocean-going life.

It may be that we will never discover enough penguin fossils to be able to reconstruct the exact evolutionary pathways they followed to become what they are today. Yet it is truly amazing to note that the oldest known penguin remains show us that as early as 60 million years ago the blueprint of the modern penguin was already well laid out. As my co-authors and I began discussions with penguin researchers around

the world, asking them to share their discoveries and experiences through their own words in a special section of this book, a kaleidoscope of arresting facts and findings began to emerge. Together, these paint a fascinating picture of how those millions of years served to hew the fine-tuned adjustments that enable penguins to routinely cross the threshold between those two drastically different worlds — land and sea — and hone their proficiency in both. In this they are vastly more advanced than any other air-breathing marine animals such as seals, turtles and even cetaceans. Because of their superlative adaptations, penguins are by far the smallest warm-blooded animals found in the Southern Ocean. They are certainly the most unlikely of all birds. 🐧

1

Stripes and Brays: The 'Jackass' Foursome

The world's most northerly penguins, the 'jackass' group (so-called because of their donkey-like braying) divides itself between South America and southern Africa, and a few islands in the vicinity of each. In addition to the classic black-and-white coloration that is the signature of nearly all penguins, they share the unusual feature of a double black-and-white band, of varying configuration, running down their face, neck and chest. A patch of bare pink skin between eye and beak, more streamlined but far less well insulated no doubt in response to their temperate habitat, provides the only touch of colour. These are some of the most easily observed of penguins, yet the threats they face create numerous conundrums for conservationists working to ensure their survival. Living closer to humans than most other species often pitches them directly into the path of many of our more pernicious activities.

Galapagos penguin

Giant cacti, sun-baked lava flows, schools of angelfish, and sea turtles grazing in the wave-wash — this is hardly the backdrop that first comes to mind as a natural habitat for penguins. Yet these are amongst my most cherished memories of childhood explorations and early photography projects during my 35 years growing up and living in the Galapagos Archipelago. I remember magic tranquil nights sleeping under the stars, the quietude punctuated with the mournful, almost melancholic calls of penguins returning to their nests. From the jagged lava headland, amorous mates would communicate with long, wistful brays — 'Aaoooo, heee-haaoooo, heee-haaoooo...' — vaguely reminiscent of a distant foghorn. For me, this represents the most evocative sound in that otherworldly environment surrounding Fernandina and western Isabela Islands, where frigid seas meet scorched lava spewed by some of the world's most active volcanoes. In this unique

ABOVE A Galapagos penguin darts along the sandy seabed in pursuit of small schooling fish, Sombrero Chino, Galapagos.
OPPOSITE Part of varied and expressive courtship rituals, a pair of Magellanic penguins stand tall while the male (slightly taller than the female) pats his mate gently with quivering flippers, Patagonia, Argentina.

GALAPAGOS PENGUIN
ABOVE Seeking to rejoin the small flock at the end of a day's fishing, a lone adult emits its mournful contact call, Elizabeth Bay, Isabela Island.

RIGHT With a puffed-up throat during courtship, a male investigates a potential nest site in a tiny lava tube, Cape Douglas, Fernandina Island.

microclimate the land is a virtual desert, whereas beneath the waves penguins share their plankton-rich undersea habitat with an unparalleled diversity of species, from marine iguanas, fur seals and flightless cormorants to giant sunfish and sperm whales.

Cape Douglas, Fernandina Island, is perfect Galapagos penguin habitat. During the mid-year change of season, thick, windless fog is a common feature of this coast where sea temperatures are often substantially colder than anywhere else along the equator. Huddled in my sleeping bag on the little beach, enveloped by the damp chill just metres from the waves lapping the coarse sand, I listen as the penguin calls draw near. Soon I sense their little shuffling feet passing right next to where I lie, quite still. In the dawn half-light, pairs are courting. They circle around each other with little hops, murmuring and nodding their heads in jerky movements, or patting each other's back and flanks with gentle taps of their stiff flippers. Soon they disappear into dark crevices and small tunnels under the

lava crust surrounding the beach. They are shy and retiring creatures, most unlike their southern relatives who are notorious for their raucous, sprawling colonies. The diminutive Galapagos penguin is the rarest and second smallest penguin in the world, standing just 35 cm (14 in) tall, and weighing roughly 2 kg (4½ lb). Only some 2000 individuals are in existence today. The Galapagos penguin could be said to represent the end of the line in penguin evolution, its ancestors shunted north all the way to the equator by the cold waters of the South Equatorial Current that flow from the Southern Ocean along western South America, further enriched in coastal areas along the way by the cold, powerful upwellings of the Humboldt Current.

Living smack beneath the equator, with a few individuals straying just north of it, the penguins of Galapagos exist thanks to a remarkable environmental quirk. Here the upwelling Cromwell Current (also known as the Equatorial Countercurrent) is felt most

ABOVE A mixed group of adults and immatures rests on the surf-beaten volcanic south coast of Isabela Island.
LEFT A wide yawn reveals the barbed tongue and palate which help grasp slippery prey. The chest and facial bands in this species are only faintly defined.

GALAPAGOS PENGUIN
ABOVE AND RIGHT Fishing is mostly done alone or in small groups, worrying and startling bottom-hugging schools of baitfish, often snatching them from below.

strongly. A thin, mid-water stream flowing west–east across the Pacific, it rises steeply from great depths along the western edge of the submarine Galapagos Platform. This surge carries deep-sea nutrients upward which, upon emerging into sunlight, nurture a veritable explosion of life at the surface — the penguins' lifeline. This allows Galapagos penguins to be coastal feeders — unlike most other penguins — rarely venturing more than 200 m (660 ft) offshore, thus the species' entire range consists of some 350 km (220 miles) of tortuous coastline, a minute habitat pocket only about 150 km long and 50 km wide (95 by 30 miles), plus a couple of other tiny footholds elsewhere in the archipelago. This represents both a minuscule population and an extraordinarily small distribution for an entire seabird species. Living at the edge of possibility for any modern penguin renders it exceedingly vulnerable to any shifts in environmental conditions.

ABOVE Diurnal and sedentary on their feeding grounds, penguins hop ashore to doze in the late afternoon sun. Able to moult any time of year, their colour variation indicates new versus faded plumage, Bartolomé Island.

LEFT A hunting penguin pops up briefly for a breath, Villamil, Isabela Island.

Humboldt penguin

Exactly 40 years after my first childhood encounter with the Galapagos penguin, I find myself once again camped on a cactus-studded desert island, but some 4000 km (2485 miles) to the south. Once more I am listening to plaintive penguin brays through a chilly, damp dawn where the sea is rich but the land is arid, the result of strong oceanic upwellings. But this time the voices ringing from the boulder shore are huskier. I am here to meet the Galapagos penguin's direct ancestor, the Humboldt penguin, whose habitat is sustained by the current of the same name. Splashing ashore through thick kelp fringes in groups of a dozen or more, they are nervous creatures, panicking easily at the sight of any movement. Perhaps for good reason, as their relationship with humans has been a long and difficult one.

In the past, their stronghold was the so-called Guano Islands along the coast of Peru, where numbers probably were in the hundreds of thousands, if not millions. They fed on vast resident shoals of anchoveta, a remarkable

small fish whose position in the food chain places it directly above phytoplankton, skipping the normal intermediary zooplankton tier. This allowed for its superlative abundance, before commercial overfishing upset the balance. On these rainless seabird islands, the penguins dug nesting burrows into centuries' worth of guano (bird droppings) accumulation.

This peaceful setting was changed forever in the 19th century. Seabird guano was discovered as a 'miracle' natural fertiliser, and 20 million tonnes was exported from Peru over a period of 40 years. The islands were literally scraped bare, robbing the penguins of their nesting habitat, and in many cases lowering the islands' elevation by 10 m (33 ft) or more. Later there came a mushrooming fishmeal industry that eventually led to the collapse of anchoveta stocks.

With no reliable food source and few remaining places to nest, their eggs often raided by fishermen and even adults and chicks poached for food or fish bait, the Peruvian population of Humboldt penguins dwindled to

ABOVE AND RIGHT Mining guano for fertiliser in the late 19th and early 20th centuries deprived the Humboldt penguin of the substrate they needed to dig burrows, so they now use rocky caves or hollows where possible, which they scratch out and line with any debris available, Tilgo Island, Chile.

mere remnants. Today they are making an attempted comeback in a few protected sanctuaries, such as Punta San Juan.

On Chile's Isla Tilgo, these reclusive penguins have it easier. Even though the island receives no official protection, as do nearby Choros and Chañaral Islands, the penguins here are doing exceptionally well at the time of my visit during the cool southern winter. Nesting in caves and fissures between boulders, or under spiny desert bushes, most are nurturing plump pairs of chicks, reflecting productive conditions in nearby waters. Commuters toddle quietly up and down well-worn pathways amongst the cacti, under the watchful eyes of hungry Turkey vultures and busy Neotropic cormorants, whose noisy young chatter incessantly in nests balanced amongst the crowns of the tallest cacti. Only a stone's throw from the southern Atacama Desert coast, it seems amazing that no predators have made it across the channel, although at night the island teems with small desert mice.

MAGELLANIC PENGUIN
ABOVE A group lands on a windswept beach in the Falkland islands.
TOP RIGHT Nesting colonies on the mainland shores of South America are usually large and much denser than on islands, perhaps to better repel land predators, Cabo Dos Bahías, Argentine Patagonia.
RIGHT Parent and chick doze peacefully at the entrance of their nesting burrow in Patagonia.
FAR RIGHT A pair mates on the shores of Seno Otway, Southern Chile.

Watching the penguins come and go through waving strands of bull kelp, preening in large groups at their landing sites, or busily feeding hungry chicks, I felt lucky, as I always do when I am amongst penguins, to share in the most intimate moments of their secretive lives.

Magellanic penguin

Lucky is also the feeling that resurged in me time and again during long camping treks along the Falkland Islands coastlines. Feeling the freedom of the cold ocean winds, my partner and I pitched our little tent in sheltered green dells every night, or among clumps of tall tussock grass, where Magellanic penguins gave full voice to nightly social concerts. Many a rainy morning we awoke to their doleful serenades echoing across the vales, as small gaggles stood together to socialise and court.

Sheltering in deep burrows to incubate their eggs and raise their chicks provides this species

with a measure of protection from predators, a luxury not available to surface-nesting penguins in more exposed habitat. Equipped with exceptionally sharp and powerful beaks, together with a feisty temper, Magellanics sitting on their nest inside the burrows can easily take on inquisitive skuas, giant petrels or other potential nest robbers, as many an 'egger' discovered in the days when the Falklands government practised National Egging Day, when schools were let out so children could gather tens of thousands of eggs that the pioneering islanders preserved as food for the winter months.

This effective defence enabled Magellanic penguins to establish themselves at a number of colonies on the continent of South America. From the Patagonian coast of Argentina, notably at Punta Tombo, down to Tierra del Fuego and beyond the Chilean fiords on the Pacific side, they nest in sometimes enormous colonies, the landscape honeycombed by their burrows. The species meets up with the southernmost nesting Humboldt penguins near Valparaiso, where I found it intriguing to

MAGELLANIC PENGUIN

PREVIOUS PAGES As the rising sun breaks through the clouds, a group of nesting adults gathers to socialise in the wave-wash for a while before heading out to feed, Saunders Island, Falklands.

BELOW Shaking their heads and clicking bills together, group courtship sessions are common, Seno Otway, Chile.

BOTTOM Digging a new burrow amongst plant roots in old glacial deposits is hard work, Beagle Channel, Argentina.

RIGHT Both greetings and courtship involve loud, drawn-out braying for which the species is renowned, West Point Island, Falklands.

BOTTOM RIGHT Once the burrow has been cleaned out, grass is carried to line the nest, Seno Otway, Chile.

see both types feeding in mixed flocks, each easily recognisable at sea by their distinctive banded collars. In the Beagle Channel, I also saw both species rubbing shoulders with Gentoo penguins, although very different choices of nesting sites meant they had little to do with each other. But regardless of where they are, the Magellanics' bold black-and-white striped flanks flashing through the surf as groups make their final dash for the beach, rapidly scuttling through the wave-wash beyond the reach of any marauding sea lions, always provides a moment of excitement.

African penguin

The very name 'African penguin' sounds somewhat oxymoronic, implausibly removed from the popular vision of penguins as Antarctic birds. The first time I watched African penguins emerge from turquoise waters and toddle up a picture-perfect, snow-white beach — complete with holiday homes perched on

ABOVE As Magellanic chicks grow older and bolder, they begin to emerge from the nest burrow to join their parents in the afternoon sun, Saunders Island, Falklands.
LEFT Mutual preening is an integral part of courtship and pair bonding, Cabo Dos Bahías, Argentina.

AFRICAN PENGUIN
ABOVE Landing on granite outcrops at Boulders Beach, an adult heads to its nest near Simonstown outside of Cape Town. Unwittingly, this residential development provides shelter from natural predators, allowing a penguin colony to flourish on the African continent.

the granite headland — it looked to me more like some sort of Penguin Club Med than a nesting colony.

In fact, the Boulders Beach nesting colony on the Indian Ocean side of South Africa's Cape of Good Hope is a recently established one that started when two pairs of penguins found a safe haven to raise their chicks between the houses and the sea in 1982. More joined them, taking advantage of the absence of wild African predators around the coastal subdivision. The colony grew to about 3000, and now attracts some 60 000 visitors a year. Until the site reached carrying capacity, this was the only colony of African penguins consistently on the increase, even while others in more natural settings dwindled. To placate residents unimpressed by their noise and smell, a fence now separates penguins from humans.

On the opposite side of the Cape, Robben Island sits just off Cape Town in the Atlantic. In spite of its dark and tumultuous history as a fortified prison camp — where Nelson Mandela spent 27 years of his

life incarcerated in his fight against apartheid — this is another African penguin stronghold. Here I watch bands of penguins returning home at sunset, just as the pretty lights of Cape Town begin to glow, framed by the impressive outline of Table Mountain against a rosy sky. Minding their own business, the penguins scuttle ashore across slippery boulders, cleverly skirting around the concrete harbour and rows of buildings, ducking along well-worn pathways through dense scrub to gain their nesting sites hidden well inland.

African penguins — variously called Black-footed or Jackass penguins — were the first ever described by modern Europeans, when in 1497 the Portuguese navigator Vasco da Gama sailed around South Africa on his way to India. But that first contact was already ominous: 'These birds, of which we slaughtered as many as we could, cried like jackass...' They are also the penguins most often seen in zoos and aquaria in the northern hemisphere, where climate similarities enable them to breed with relative ease in captivity.

However, all does not bode well for their future. From an estimated original population of around 1.5 million a century ago, today a mere 25000 breeding pairs remain, and even these numbers are still dropping. In the early days of human settlement the penguins were liberally consumed for food. Then coastal development made further inroads, followed by massive oils spills from ever-larger numbers of modern supertankers plying the notoriously stormy seas around the bottom of Africa.

To make matters worse, since entering the 21st century, the sardine runs upon which these penguins depend have been moving gradually farther south, apparently in response to global warming and fisheries pressures. This causes massive breeding failures when parents are unable to cover the distances needed to keep their chicks well fed. Whether our best-known penguin will make it into the 22nd century is a question that only time will answer.

ABOVE With speckled breast and dark feet that also earned it the name of Black-footed penguin, this was the very first species described by European seafarers in 1497, noting its call as similar to that of a donkey, or jackass.
LEFT A pair seeks the shade of a boulder to preen, Boulders Beach, South Africa.

2
Long-tailed Trilogy: The Antarctic Trio

The height of summer on the south coast of King George Island, in the South Shetlands, doesn't feel all that much like real Antarctica. In this maritime region of the Antarctic Peninsula, temperatures at this time of year tend to stay well above freezing, and even the annual mean hovers not far below 0°C. This is home to all three species of long-tailed penguins — the true Antarctic penguins — representing the entire genus *Pygoscelis*. But even within this wildlife-rich area, very few locations can boast all three types nesting alongside one another; their specific feeding grounds and habitat preferences usually segregate their choices of nest sites.

Where similarities end

Having shed my parka and several layers of warm clothing, I lie sprawled on the smooth, sun-warmed beach pebbles, taking in my surroundings. Remnant patches of snow cling to a few gullies and southfacing hillsides, while the flatter areas near the shore, wet with meltwater, have turned vibrant green, like freshly

sprouted lawns. In actual fact, there is no grass here, just fast-growing ground alga thriving on guano run-off (a single species of Antarctic hair grass found in the Peninsula region grows only in small clumps in sheltered rocky areas). Even the snowbanks have acquired surprising hues under near-constant daylight: yellowish, brownish and even bright pink in places, the work of other microscopic algae, single-celled diatoms commonly referred to as ice algae. Still brighter are the colours added by gaudy splashes of flame-orange lichens encrusting the faces of pillar-like volcanic rocks. What drives much of these flourishing concentrations of life, apart from the relatively mild climate, is of course the penguins.

Because of its exceptional density and variety of life, this place has been set aside as SSSI #8 under Antarctic Treaty regulations, that is, a Site of Special Scientific Interest, where access is limited and great care must be taken not to damage the delicate living communities. Despite this being adjacent to one of the most active

ABOVE Near the southern limit of their range, a pair of Gentoo penguins nesting on the continental shores of the Antarctic Peninsula hunker down on their nest during a blizzard, Brown Bluff, Antarctic Sound.
OPPOSITE A tightly packed colony of Adélie penguins helps colour the landscape on King George Island in the South Shetland group, their krill-tinged guano nurturing life through algae and lichen growth after the snowmelt.

ABOVE Faithfully returning to their tried and tested nest sites, early in the season Gentoos commute up well-marked, krill-tinted pathways to reach snowfree rocky ridges, Neko Harbour, Antarctic Peninsula.

RIGHT (ALL THREE) The respective ranges of the three brush-tailed penguins — Adélie, Chinstrap and Gentoo — overlap only around the Antarctic Peninsula, where their breeding habits reflect differing habitat preferences.

continuously running research stations in the region — Arctowski, established in 1977 by the prestigious Polish Polar Academy — the area remains well-preserved, with human activity generally restricted to research.

Observing the penguins' busy comings and goings, it doesn't take me long to notice patterns of behaviour that reflect their distinct lifestyles: the three Pygoscelid penguins respond in quite different ways to both the opportunities and limitations imposed by sea-ice.

Chinstrap penguins are continuously toddling up and down fairly steep, rocky slopes, tending delicate pale grey chicks hatched only a week or two ago, as both parents frequently swap roles between brooding the chicks for protection and warmth, and provisioning trips to provide for their rapid development. In startling contrast, the Adélie chicks are already well on their way toward independence, plump but agile balls of

sooty-coloured down wandering about the colonies, which are centred on the lower flats. Already three-quarters grown, they will depart the land in just three to four more weeks, around mid-February. Closest to the shore are the Gentoos, gaggles of large, fat chicks gathered in same-age 'crèches' while their parents are away searching for dinner. Some smaller Gentoo chicks remain nest-bound, with doting adults sheltering them from inclement weather and ever-vigilant, hungry skuas. Unlike their neighbours, the Gentoos seem relaxed, moving about unhurriedly. Two more months still lie ahead of them before their chicks too will become independent.

It is late January, the warmest part of summer, but even though all three penguin species have chicks, their respective stages of development are indicative of different Antarctic environments for which each is best adapted. This explains why there are so few areas where the three species actually converge.

ABOVE Gleaming in their dense, watertight plumage, Gentoos rush back to the colony at the onset of the nesting season, Cuverville Island, Antarctic Peninsula.

Adélie penguin

With frequent shudders and loud crunching sounds, the Australian icebreaker *Aurora Australis* is forging its way, metre by metre, through thick fast-ice, the kind of sea-ice that is welded to the shore in a solid, often impenetrable, expanse. Sixteen days out from Hobart, Tasmania, and over 5000 km (3100 miles) as the petrel flies, we are finally within sight of land. A dark strip of rocky coastline with a blurry cluster of brightly painted buildings, barely visible through a rising blizzard, heralds our arrival at Davis Station, one of three Australian research centres in East Antarctica. A strong sense of expectation is spreading among the 80-plus expeditioners on board. For the geologists, physicists, meteorologists, biologists, along with a bevy of support personnel about to be dropped here, including carpenters, plumbers, electricians, cooks, diesel mechanics and many other professionals, Antarctica beckons. Their adventure is about to begin. But for me, at least part of my objective was already achieved 10

days ago, upon sighting the first Adélie penguin when we left the open ocean and entered the pack-ice, still 600 km (375 miles) from land.

As a guest of the Australian Antarctic Division, I feel exceptionally privileged to be approaching Antarctica in early November, ahead of the sea-ice break-up and well before most visitors reach the continent. We have travelled south in tandem with the Adélies. They are making their spring bid to reach land to begin nesting, and their numbers increase daily as they, too, negotiate the still-frozen sea. Tiny black dots in a vast crystal world, they resemble long lines of ants in the distance, scrambling across broken pressure ridges, then flopping onto their bellies to glide effortlessly over flatter ice pans. As daylight fades into the twilight of the polar night, soft hues reflected from massive trapped icebergs shift from deep turquoise to violet-purple, creating an ethereal scene.

Like us, the penguins navigate unerringly toward their familiar destination. But unlike us, they make no

ADÉLIE PENGUIN

ABOVE With egg-laying already underway at the colony, a small group hurries back to the distant open water to feed, crossing several kilomtres of frozen sea in the spring twilight; at about the time the sun stops setting, the winter ice will break up, shortening their commute during the brief summer in Prydz Bay, East Antarctica.

RIGHT Penguin toenails act as effective crampons on wind-polished sea ice near Gardner Island, East Antarctica.

use of radar and GPS to know precisely where they are headed. A feeling of both great energy and vital urgency emanates from their eager march, frantic to reach the islands weeks ahead of the summer thaw. This timing will allow them easier access to the sea when chicks need feeding. Indeed, natural selection has built into their senses the need to complete the nesting cycle within the shortest window of opportunity before the return of deep-freeze conditions, and that means an extra early start in this extreme part of the planet.

A few days later, in the company of a couple of researchers, I walk across the ice to reach one of the busy nesting islands just a couple of kilometres from where our ship is now firmly 'parked', busily offloading a whole year's worth of fuel and supplies for Davis Station. The noise of heavy machinery, cargo cranes and trucks rumbling across the glistening sea-ice recedes as we walk, replaced by the soft squeaking of our boot chains biting into the frozen surface. Two months from now, waves will be lapping the beach here.

Soon a small band of curious Adélies are catching up with us, craning their necks quizzically for a close inspection, waddling alongside for a while. Prospectors still too young to start nesting, but already feeling the pull of colony life, they resume their urgent explorations and quickly disappear into the blue-grey distance.

On the island, matters are far more earnest. Territories are guarded and defended, old nest scrapes are being scratched and reshaped, and pebbles for their lining pulled with difficulty from the still-frozen ground. Standing tall, their beaks pointed to the sky, expectant males awaiting the return of their mates sing out loudly, pumping their flippers slowly for better effect. Pairs everywhere are being reunited in an explosion of screams and head waving known as the 'ecstatic display'. But strife erupts equally abruptly when a returning female discovers another has already won the attention of her mate of last season.

Males have normally arrived a week or two ahead of the females, to lay claim to their tried-and-tested

ABOVE In the vastness of the Weddell Sea, a lone Adélie rests on an ice floe.

LEFT A film of water adds gloss to a porpoising Adélie's streamlined shape rising for a breath from a windless sea, Antarctic Sound.

FOLLOWING PAGES Cape Adare in the Ross Sea holds the largest Adélie colony, with about a quarter of a million nests occupying the high grounds, while commuters splash across meltwater pools at the end of summer.

ADÉLIE PENGUIN

TOP LEFT Parents examine their new hatchlings, King George Island, South Shetlands.

TOP RIGHT The midnight sun caresses nesting colonies on Gardner Island, East Antarctica.

ABOVE LEFT Collecting pebbles for nest lining, Devil Island, Weddell Sea.

ABOVE RIGHT Splendid mountains overlook colonies on Petermann Island, Antarctic Peninsula.

RIGHT A male advertises his search for a mate on Gardner Island, East Antarctica.

nest sites of prior years. Most often, their previous mates rejoin them here, where they quickly settle down together for the new nesting cycle. But with the urgency to begin breeding without wasting time, males can't afford to wait for long. It is when a female returns to find another has already usurped her place that — perhaps unsurprisingly — the most vicious and vociferous battles break out. Bites, kicks and bone-hard flipper spanking in lightning-speed volleys are all part of the 'no-holds-barred' tussles. Blood, mud and feathers fly as they tear at each other with unbelievable ferocity. Blindly, they tumble through the colony while showering each other in punishing blows. Ripples of indignation spread as they knock neighbours off their nests, until one makes an escape with her nemesis in hot pursuit. The speed of the battle is such that it is only later, when reviewing my (mostly blurred) photos that I discover that, besides pounding flippers and jabbing, twisting dagger-like bills, they also lash out with kung-fu-style kicks to topple their opponent.

An hour before midnight the sun sinks past a blood-red slit in the clouds, angling slowly beneath the horizon for a couple of hours of delightful twilight. Just then I discover the first few eggs have already been laid; pale aqua in colour and still spotlessly clean. By the time they hatch, in about five weeks, the sun will have stopped setting altogether.

ABOVE As winter approaches and penguins prepare to leave, evening mist rolls over the Cape Hallet colony in the Ross Sea.

LEFT Their stomachs bulging with food for their growing chicks, parents hurry across a snowmelt pool at Cape Adare, Ross Sea.

FAR LEFT A well-worn pathway leads to shore across a rampart of broken sea ice, Possession Island, Ross Sea.

Chinstrap penguin

CHINSTRAP PENGUIN
ABOVE Painting the
volcanic landscape of
Deception Island in subtle
hues, pink krill-stained
guano contrasts with a
flush of green ground alga
profiting from the leached
nutrients around the huge
colony at Bailey Head.
RIGHT The remarkable
facial design of this
species is unique among
penguins, Half Moon
Island, South Shetlands.

The Antarctic Peninsula, and especially its surrounding
islands, is at the heart of Chinstrap territory. Their
preferred feeding habitat is along the outer edge of
seasonally pulsating pack-ice where it is patchy and
scattered. In summer, flocks can sometimes be seen
resting on spectacularly weathered blue-and-green
icebergs rocking in the swells — the last remnants of
compact glacial ice drifting in the open ocean.

While it is impossible to have one 'favourite' penguin,
the Chinstrap's slender body and lithe demeanour
and its decorative chin-band make it one of my top
contenders. But I'm not referring to one Chinstrap
in isolation, rather to the combined effect of seeing
thousands of them all at once, flowing up and down
valleys, squabbling and calling and generally going
about the busy business of raising chicks and defending
nests. I doubt a better place exists to appreciate
Chinstrap penguins in all their glory than Bailey Head
on the eastern shore of Deception Island. In a good

LEFT Hatching paler than all other penguins, the first down of Chinstrap chicks is silvery grey, Nelson Island, South Shetlands.

ABOVE The rugged, lichen-clad slopes of Bailey Head's old volcano offer shelter to a well-fed family, Deception Island.

BELOW Waving his flippers and pumping his chest, a sky-pointing male advertises his desire for a mate, Half Moon Island, South Shetlands.

season over 200 000 pairs line the dark, foreboding slopes of an imposing volcanic amphitheatre, their mingling voices echoing off the surrounding walls and rising on the wind. In coordinated waves they splash ashore between frothing breakers, their white breasts sparkling on the black scoria beach beneath an equally black, ash-layered glacier carrying the records of the island's recent eruptions. After a perfunctory shake and preen, the throngs trundle up the only access valley, paddling waist-deep up gurgling meltwater rivulets, then fanning out and upward, each to his or her nest, while just as many are making the opposite journey back down.

I join the convoys and make my way slowly to higher ground, stopping often so as not to obstruct two-way traffic through narrow passages. Every rise and knoll, the first areas to lose their snow load in spring, is carpeted in evenly spaced penguin nests. While heavy mist hangs over the glaciated ridges above, tendrils of warm sunshine raise swirls of

CHINSTRAP PENGUIN
ABOVE The dense blue, wave-polished icebergs, slowly dying at sea beyond the edge of seasonal pack ice, present resting opportunities for penguins on foraging trips from the large colonies in the South Orkney Islands, Scotia Sea.

steam from the dark pink-and-green tinged volcanic ash. When I reach the crater's edge, I find star-like guano 'artwork' surrounds each nest, where pearly grey chicks snuggle under protective parental breasts. Far below the volcanic brim, the view drops sheer to the shoreline and the pale blue rippled surface of the Bransfield Strait, barely visible under a gauze-like sea fog.

While every penguin island is a marvel to behold, perhaps the most awe-inspiring of all are the South Sandwich Islands in the farthest reaches of the South Atlantic — Chinstrap Grand Central. At Zavodovski Island, the largest Chinstrap colony in the world, said to number over one million pairs, swaddles the sombre flanks of an active volcano, the 551-m (1808-ft) Mount Curry, lost in dark roiling clouds laced with sulphuric fumes. I visited just two others of the 11 islands that make up this young volcanic chain, all but one home to sprawling Chinstrap colonies.

Chinstrap penguins also surrounded Antarctic explorer Sir Ernest Shackleton's men on unimaginably

rugged Elephant Island, where they survived on not much else while awaiting rescue in 1916. Watching storm clouds tumble over hanging glaciers, and hearing the Drake Passage swells crash on the foreboding shore, I find it difficult to consider that penguins can live and breed here. Much less can I imagine 21 shivering men huddled under an upturned longboat for nearly five months. Even though a good many penguins were boiled for sustenance on their flickering blubber stove, perhaps their antics also helped keep the desperate men sane.

Gentoo penguin

A remarkable generalist amongst a family of specialists, and third largest of all penguins, the Gentoo feeds on pretty much any prey that comes along, from crustaceans to squid, bottom fish to free-swimming Antarctic krill. It has also adapted to a wider range of habitat conditions than any other penguin, split into two slightly different races: the southern subspecies as

ABOVE Under the face of a glacier stained black by past volcanic eruptions, Chinstrap commuters come and go on the scoria beach at Bailey Head, Deception Island. LEFT Perhaps looking for a secluded place to moult, a fat Chinstrap wanders through curtains of steam on the beach of the active caldera of Deception Island.

GENTOO PENGUIN

LEFT (ALL THREE) In just a few weeks' time chicks develop from tiny vulnerable hatchlings to mobile and inquisitive, gangly youngsters getting ready to moult into their first coat of feathers, running about through mud and puddles, exercising their wings and squeaking constantly, Falkland Islands.
BELOW Young adults on the flooded outskirts of the colony court and practise nest building for seasons to come — including a fake pebble-dropping gesture, Hope Harbour, West Falkland.

much at ease on ice-clad shores around the Peninsula as its slightly slimmer northern counterpart is on windswept subantarctic islands north of the oceanic convergence. Unlike its closest relatives, it is not a species associated with sea-ice, even in the southern parts of its range. Instead, it is a remarkably deep diver, and has been recorded reaching depths over 200 m (656 ft), making use of areas of deep water close to land. Rather unusual for a penguin of Antarctic latitudes, it is mainly sedentary, with small colonies concentrated where strong tidal currents maintain open water throughout winter. These traits also mean that the Gentoo's timescale for breeding is not as constrained as it is for other Antarctic species, explaining why eggs may only just be hatching while nearby Adélie chicks are already preparing to fledge.

I find an air of endearing nonchalance surrounds Gentoos. Their life seems gentle, unhurried and well organised compared to the hell-for-leather lifestyle of their smaller brethren. They are adapted to the fringes

of the Antarctic, whereas Adélies are most at home in the much harsher conditions of the far south, sharing little range overlap.

The markings and colours of Gentoo penguins are both intense and distinguished to the human eye, and their slow, rhythmic calls sound peaceful and slightly melancholic. All around the Peninsula, from Petermann Island and Paradise Bay to many different spots along the South Shetland and South Orkney Islands, my memories of Gentoos are punctuated with tranquil moments. On many occasions, toddling chicks have wandered up to me inquisitively, pulling on my trousers and sleeves, and on several occasions have even fallen asleep trustingly nestled against my legs as I sat motionless on the ground.

Gentoo penguins nesting in the Falkland Islands present a different impression, although their pace is also relaxed. Instead of the traditional pebble nests they build on Antarctic shores, here they strip vegetation to build large mounded nests of twigs and peaty sod.

Much-feared Gentoo predators include (LEFT) bull sea lions taking adults, (BELOW) skuas swooping down on young chicks, and (BOTTOM) Striated caracara stealing eggs, Hope Harbour, West Falkland.

With each new nesting season they choose different locations to establish dense colonies amongst carpets of low plants, sometimes far up valleys or hills more than a kilometre (over half a mile) inland. Trampling the coarser groundcover, their commuter lanes can be recognised from afar as green avenues wending up from the shore, the windswept turf fertilised by their rich guano.

Remarkably resilient, I've seen thriving Gentoo colonies in many contrasting environments. In the Beagle Channel near the tip of South America they were panting in the heat of a Patagonian summer's day, while at South Georgia I watched them hold their own among aggressive fur seals. I found them half-buried in a Weddell Sea blizzard while incubating their eggs at Brown Bluff, and on the opposite side of the continent, they were strolling amongst elephant seals on Australia's isolated Macquarie Island. Invariably, their way of going about life leaves me with a reassuring impression of serenity.

Antarctic circumnavigation

During the 1990s, I used to work as Expedition Leader on a powerful icebreaker chartered for passenger trips to Antarctica. Of these, the most extraordinary voyage I ever led was the first-ever complete passenger circumnavigation of the Antarctic continent — two months of exploration with the only certainties being our start and end dates at the Falkland Islands. The rest was up to me and the captain, and, most notably, the ice master who, with the help of our three capable helicopter pilots, would see us into, through and around the many unforeseeable ice barriers along the way. We sailed from Port Stanley on 25 November 1996 and returned on 26 January 1997.

What I hadn't realised before we set off was that in one totally amazing continental-scale clockwise spin, we would span not only the entire Antarctic summer, but with it bear witness to the frantic pace of the Adélie penguins' complete nesting cycle, leaving the less ice-hardened species in our wake.

Three days into the trip we made our first landfall at the South Orkney Islands in the far South Atlantic, an outer bastion of the Weddell Sea. The entire colony was unusually quiet, saving energy for the 33 days of incubation just begun. Three weeks and a full quarter of the way around the continent later, we found Adélies still sitting tight at Scullin Monolith, a gigantic crescent-shaped granitic extrusion rising dramatically out of the sea at the edge of the Antarctic icecap. In a past season a cataclysmic avalanche must have tumbled down its polished grey palisades, overwhelming much of the nesting colony. The survivors now hunkered grimly amid mummified, freeze-dried victims left withering in the desiccating climate.

One month into our trip, we reached Gardner and Peterson Islands almost diametrically opposite our first Adélie encounter. With Christmas upon us and the midsummer sun perpetually high above the horizon, sooty-black chicks were literally popping out of their eggs everywhere, feeble peeps and small wobbly heads

begging frantically to be fed. It is a tight race against time and the Adélies' pace suddenly switches into high gear. For the parents, that race means a frenetic scramble to provide sufficient food for transformation from tiny helpless chick to fully feathered fledgling possible within the excruciatingly short span of six to nine weeks.

On day 45 we reached Cape Royds deep in the Ross Sea, the southernmost of all present-day Adélie colonies. Here we found the chicks about half as large as their parents and already beginning to grow patches of waterproof feathers to replace their baby down. Clearly, at 77°33' S latitude, a penguin's imperative is loud and clear: fledge fast or die.

Two weeks later, on 23 January, we were back in the South Shetland Islands bordering the western side of the Antarctic Peninsula. With just three days left to complete our voyage, the time had come to say goodbye to the penguins — and they in turn were saying goodbye to land. No longer concerned with staying near their nest sites, plump, almost fully feathered chicks wandered about, funny tufts of remnant down giving them Mohawk-style hairdos. With that wide-eyed look that their developing white eye-rings give them, they stared toward the sea as though attracted and frightened at the same time. Within days, the colony would be silent once again as the dark grip of another winter took hold.

GENTOO PENGUIN

ABOVE Returning from foraging, parents caring for chicks storm the beach in huge flocks, seeking safety in numbers from marauding sea lions, Volunteer Beach, East Falkland.

RIGHT (BOTH) A case of perfect mis-timing results in a mid-air collision between arriving and departing birds, Death Cove, Hope Harbour, West Falkland.

ABOVE Commuters wend their way over dunes covered in blooming sea cabbage, Volunteer Beach, East Falkland.
LEFT AND BELOW Returning adults explode from the water in fear of a sea lion ambush, sometimes timing their approach to ride the crest of a wave, Death Cove, Hope Harbour, West Falkland.

3
Island Dandies: The Crested Penguins

With a total of seven species amongst their ranks, the crested penguins are an unusual bunch. All sport extravagant golden head plumes of various sorts, some fanning out from the forehead and trailing smoothly aft (Macaroni and Royal), others growing into neat bushy eyebrows (Snares and Fiordland), and one boasting the distinction of stiff, raisable bristles in two rows jutting straight up punk-like (Erect-crested). There are two more penguins (Southern and Northern Rockhoppers, which I'll treat in their own chapter) whose comical headgear seems in constant disarray, at times worn as an elegant crown and at others as wild dreadlocks, like a penguin version of a really bad hair day.

Erect-crested penguin

The Erect-crested penguin is one of the least known of the *Eudyptes*, or 'crested', tribe — flamboyant little penguins that live on just two of the remotest, storm-lashed island groups in the subantarctic seas south of New Zealand. Just to see one is a rare privilege. That's

what I remind myself of in the small hours of the night, as ear-piercingly loud, staccato squawks incessantly assail my ears from both sides of the tent. It is mid-November and chicks are about to start hatching on the Bounty Islands. Night and day, thousands of incubating parents are responding to the tiny peeps emanating from within the eggs by arching their heads down and calling rhythmically every few minutes to imprint their voices for later recognition. I snuggle contentedly into my sleeping bag and try to doze. The wind is howling overhead, and giant swells thud against the cliff face below. Our meagre campsite, wedged between nests, is perched at a crazy angle on a sloping granite slab — it's as wild as wild can be. I'm in penguin heaven.

About 650 km (405 miles) southeast of New Zealand's South Island, the Bounties, together with the Antipodes some 220 km (140 miles) further south, are home to the world's estimated 67 000 breeding pairs of Erect-crested penguins. A collection of 13 bare granite rocks barely emerging from the fierce wavelash

ABOVE Erect-crested penguins strengthen their pair bond with prolonged mutual preening, Antipodes Island, New Zealand Subantarctic. LEFT Snares penguins dry off on the shore before heading inland to their colony hidden in the tangled *Olearia* forest, Snares Islands, New Zealand Subantarctic.

ERECT-CRESTED

ABOVE A sudden squall sweeps over the colony, Antipodes Island.

RIGHT The huge Orde Lees colony on the east coast of Antipodes Island.

FAR RIGHT *Mahalia* at Proclamation Island during our 2004 Bounty Island expedition.

OPPOSITE TOP LEFT Nesting space is tight on Bounty Islands, shared with Salvin's albatross and Fulmar prions.

OPPOSITE TOP RIGHT This species can raise its bristly crest at will.

of the Southern Ocean, the islands were discovered by Captain William Bligh in 1788 and named after his ship, HMS *Bounty*, just months before the infamous mutiny that made history in the South Seas. They are without a doubt the least visited place in New Zealand's subantarctic region, today a World Heritage Site consisting of five widely dispersed island groups, rigorously protected as National Nature Reserves by the Department of Conservation. With a combined total area of just 1.35 km² (half a square mile), every centimetre of dry surface in the Bounties is coveted as breeding territory by a mixture of penguins, albatross, petrels and fur seals.

In 2004, we mounted a private Bounty Island expedition aboard our 13-m (43-ft) sailboat *Mahalia*, skippered by my co-author and photo partner Mark Jones. With three volunteers to help collect data for several research projects, I was able to stay ashore while Mark dodged storms and kept *Mahalia* safe by ducking into wave-washed nooks among fearsome-

looking sea stacks. We were the first people to land here in seven years, and only the third party to live ashore in well over a century; the last fur seal hunters scoured the islands for pelts in the 1880s.

One of our team was Jacinda Amey, here to conduct a repeat albatross and penguin census for comparison with results from a prior expedition. Using identical methods, her absolute count of 2788 nests on Proclamation Island, where we were camped, was just 14 below her 1997 figure, an insignificant drop statistically speaking. This provided reassuring evidence that at least on this remote outpost, the Erect-crested penguin population has remained stable in recent times, unlike related species elsewhere.

With personal space at a premium in the crowded colonies, the feisty Erect-crested penguins are the staunchest defenders of their precious nest sites, around which much larger albatross and fur seals steer clear of their jabbing beaks and flailing flippers. Yet at the strange sight of humans their curiosity very

ERECT-CRESTED

LEFT Deep grooves scour the rocks where untold generations of penguins have travelled up and down steep volcanic ledges to reach their crowded colony — the largest for the species — near Orde Lees Islet, Antipodes Island.

often got the better of them, hopping closer for careful scrutiny, including investigative nips at my gloved fingers, boots and leggings. My time on the Bounties left me with many cherished, indelible memories of this isolated, thriving seabird world.

Later, on Antipodes Island, I visited one of the largest Erect-crested colonies of all. So densely packed was their vibrant metropolis that I could not insinuate myself into their midst to take advantage of easy walking where multitudinous little penguin feet had trampled the ground bare. Instead, I was forced to navigate through thick, waist-high tussock grass, observing their boisterous activities from the periphery. Earlier that day I had passed old rotting bales of penguin skins under a dank overhang in the landing cove, still rolled up where sealers more than a century ago had stashed them for pick-up by a ship that never came. In contrast, it was a pleasure to see that here, at least, penguin life had improved vastly from those days when humans pillaged even this far-flung corner of the planet.

ABOVE About one
month after the end
of the nesting season,
penguins return to their
colony sites on Snares
Islands for their annual
moult, often perching on
tree stumps in the forest
to stay clear of the
muddy ground below.

Snares penguin

Crested penguins seem to have a propensity to isolate themselves on the remotest islands, perhaps none more so than the Snares crested penguin whose breeding is restricted to the islands of the same name, about 100 km (60 miles) south of Stewart Island — New Zealand's southernmost human settlement.

With just 3.3 km² (1¼ square miles) of steep forested land to call home, some 30 000 breeding pairs return here each spring. Scrabbling up steep granitic ramparts and deep muddy gullies they promptly melt into the dark, dense understorey of two types of tree daisies, *Olearia lyalli* and *Brachyglottis stewartiae*. From their kelp-fringed landing spots, I follow their trails inland to find many small colonies tucked away in sodden, slimy mud hollows. The going is tricky, bending and twisting between the intertwined, gnarly trunks, slipping and sliding on the grease-like black peaty ground riddled with millions (literally) of Sooty shearwater burrows, trying hard to avoid collapsing them. Suddenly a screechy

scuffle draws my attention to two penguins arguing over a prime perch on a sprawling tree trunk almost 2 m (6 ft) off the ground. I do a double-take... am I really seeing penguins in trees?

The rest of the colony seems as happy as 'pigs in muck', pitter-pattering through the quagmire, their white aprons liberally smeared. The whole setting seems oddly unpenguin-like, but no doubt shelter from both vicious storms and swooping skuas brings considerable benefits. They share their leprechaun world with colourful Buller's albatross, who also nest among convoluted tree trunks and the giant leaves of rhubarb-like *Stilbocarpa* plants. As dusk falls, Sooty shearwaters darken the skies in their millions, with eerie calls soon making the entire island resonate as if it were itself a lowing animal. Other petrels join in more quietly under the cover of darkness: Diving petrels buzzing in like avian bumble-bees, and various prions fluttering moth-like around headlands. From my perspective, this is not at all a bad choice for a penguin's only home.

LEFT The main penguin landing slip is situated along a vertiginously steep granite slope, where they must first negotiate the swirling beds of bull kelp before scrambling up the smooth face leading to the forest above.

SNARES PENGUIN
ABOVE Streaming out of the *Olearia* forest on Snares Islands, a large group heads for the sea, past sleeping fur seals and through the shoreline kelp fringe.
FAR LEFT Peeking shyly from under fern fronds in the forest.
LEFT Rain pools in the shoreline granite afford drinking spots to penguins passing by.

BELOW The ecstatic greeting display between mates is loud and raucous, with strident braying and head-shaking.

BOTTOM (BOTH) A tidepool on the way to the sea is an attractive place to clean muddy feathers in vigorous bathing sessions.

Fiordland penguin

The Snares penguin's closest relative, the Fiordland crested penguin, has similar nesting habits, but is only found on mainland New Zealand. There are in fact few places on earth that feel more primeval and mysterious than the dark, fern-filled, moss-clad forests of New Zealand's South Island, especially along its southwestern coast, known simply as Fiordland. Based on fossil evidence, this forest has changed little from the time when it was still a part of the supercontinent Gondwana 80 million years ago and dinosaurs roamed in its glades. This is a land where not too many centuries ago giant eagles swooped in on 3-m (10-ft) wings to knock down monstrous flightless birds known collectively as moa. The largest of the 11 moa species reconstructed from fossil bones lived in somewhat more open country in the lee of the mountains, but even at 3.6 m (12 ft) high, and weighing up to a quarter tonne, it was no match for the 15-kg (35-lb) eagle, the largest raptor ever known.

No wonder then, that the Fiordland penguin — who outlived them all — is a shy and secretive creature. Slinking in and out of this same forest, whose fringes overhang the shoreline, the penguins arrive under the cover of twilight to nest in small aggregations amid the densest thickets they can find. Even after seeing returning parents scuttle nervously from the wave edge, and watching them from a safe distance disappear up small streambeds, it took me considerable time to locate their destination. Crawling silently on hands and knees, listening carefully for long moments, I caught the faint peep of a begging chick. This, and the abrupt squawk of squabbling adults, helped to guide me. Slithering onward, I eventually spied several nests tucked underneath dead logs and knotted roots, where tangled vines allowed almost no daylight to penetrate. Adults were busy feeding chicks and preening one another, for a long time quite unaware of my presence. When I did slip, causing a twig to snap underfoot, they all looked up

FIORDLAND PENGUIN
ABOVE Near the southern end of the species range on Codfish Island, a secretive small colony is tucked deep beneath dripping fern fronds.
RIGHT Returning to its nest late in the afternoon, an adult walks purposefully up Murphy's Beach, Westland.

fearfully, quickly backing into the darkest recesses — in stark-contrast to their arrogant cousins on the Bounty Islands.

The Fiordland penguin has picked a rather larger than usual island — larger than England — for its only home, and this has caused it considerable problems. When the first Polynesians, ancestors of today's Maori, reached these islands somewhere between 800 and 1200 years ago, penguins soon became a staple part of their diet. This led some species to extinction and shrunk the Fiordland penguin's range considerably. Then came the Europeans, who, along with their farm animals, brought predatory mammals, such as cats, pigs, stoats and other mustelids, to a country which originally had only two species of bats as sole terrestrial mammalians. It is therefore a remarkable credit to this shy penguin's tenacity that it survives at all, with an estimated 2500—3000 breeding pairs hanging on to deep forest enclaves in the most inaccessible areas.

LEFT Hidden quietly in dense forest undergrowth, growing chicks await the return of their parents, either alone or in small groups, Westland.
FAR LEFT For their first two or three weeks, chicks are brooded by the male (at left), while the female alone brings food, Jackson Head, South Island, New Zealand.
BELOW Boulders provide shelter for an unusually open nesting group at Jackson Head.

ROYAL PENGUIN
ABOVE Busy traffic crowds
the landing beach at
Sandy Bay, Macquarie
Island, marked by
constant bickering.

Royal penguin

Australian Macquarie Island, about 1500 km (930
miles) southeast of Tasmania and 1000 km (620 miles)
southwest of New Zealand, is a classic subantarctic
island. Green with tussock grass and shimmering
giant megaherbs, but with no trees growing in the
near-constant westerly storm winds, it is the throbbing
hub for numerous marine birds and mammals. Light-
mantled albatross swing their aerial courtship across
its grey skies and Southern elephant seals pile up on
its beaches in steaming mounds to sleep, mate, fight,
pup, nurse, wallow and moult. During the summer, the
sea adjacent to Macquarie's shores literally boils with
splashing hordes of King penguins. And true to crested
penguin predilection, the Royal penguin makes this
island its only home, to the tune of some 850 000 pairs.

With a maximum weight of about 8 kg (17½ lb)
and 70 cm (2¼ ft) in length, this is a hefty penguin by
crested standards, the largest of them all. It is also one
of the most gregarious, as well as pugnacious, afraid

of nothing and constantly picking fights in the midst
of its incredibly dense-packed colonies. Even moulting
individuals standing around the beaches find reason to
scrap amongst themselves, or pick on the much larger
and peaceable King penguins.

More than once I've been reduced to uncontrollable
giggles as gaggles of Royal penguins surround me;
either trooping by inquisitively on their commute to the
colony, or pecking, jabbing and tugging at my boots
and clothing with insatiable curiosity as I lie back on
the warm black sand — some of my best penguin
memories ever.

Macaroni penguin

In every crowd there have to be some non-conformists.
The Macaroni penguin is the most notable exception
to the island-specific preferences of many crested
penguins, nesting on some 50 islands and islets
stretching between the tip of South America and the
Antarctic Peninsula to the southern Indian Ocean. On

ROYAL PENGUIN
LEFT A pair of subadults shows the variability of the grey cheeks and throat that normally pale with age, as well as the massive bill and prominent pink gape typical of the species. BELOW Arguing among neighbours appears to be a compulsive occupation. BELOW LEFT Nesting colonies are so densely packed that no plants survive, creating muddy openings in the dense tussock grass. All photos Macquarie Island, Australian Subantarctic.

MACARONI PENGUIN
ABOVE Segregating from the abundant Chinstrap penguins on Candlemas Island, in the rarely visited South Sandwich group, substantial numbers nesting on the smouldering volcanic chain have recently been confirmed by satellite photography.
RIGHT Ever argumentative, as are all crested penguins, a squabble erupts between a Macaroni and a neighbouring Chinstrap, Livingston Island, South Shetland group.

five of these island groups, Macaroni numbers run in excess of one million pairs (South Georgia, Crozet, Heard, McDonald and Kerguelen). The world population mounts to an estimated 9 million breeding pairs, plus at least as many pre-breeders less than five years of age, making the Macaroni the most numerous of all penguin species.

I have seen them equally at ease scrambling up steep, wave-pounded rock slabs at Elephant Island as negotiating deep snow on Candlemas Island in the remote South Sandwich group. With a tendency to wander far and wide, occasionally a few may settle amongst other species to satisfy their gregarious nature far from their own colonies. For example, I have watched pairs holding their own amongst taller Chinstrap penguins in Antarctica's South Shetland Islands, while others in the Falkland Islands dominated much smaller Southern Rockhoppers. On Kidney Island, a lone Macaroni paired up with a Rockhopper had produced a chick that was intermediate in shape

and size — a 'Rockaroni' — and presumably infertile. Multitudinous numbers do not, however, equate to ease of access. Though I have come temptingly close on several occasions, visiting a true Macaroni megalopolis is a dream that so far eludes me.

ABOVE LEFT Agile climbers, Macaronis seem to enjoy perching on steep rocks, Hercules Bay, South Georgia.

ABOVE RIGHT The name Macaroni was inspired by the plumed headgear of 18th-century upperclass dandies, South Shetland Islands.

LEFT During the first third of the incubation period, both parents stay at the nest, taking turns sitting, Hannah Point, Livingston Island, South Shetlands.

4
Multitudes in Deep Trouble: The Rockhoppers

Anyone who's ever met a Rockhopper penguin — or better still, 10 000 Rockhoppers together, all coming and going, hopping and squabbling, courting and preening, and of course screeching at the top of their lungs — cannot fail to be seduced by their effervescent character, their sheer energy and pluck in the face of often overwhelming odds. They are the smallest of all Southern Ocean penguins and, with a body weight of around 2.5 to 3 kg (5½–6½ lb), are in fact the smallest of any warm-blooded animal spending months on end submerged in such frigid seas.

Everything Rockhoppers do they undertake with purpose and determination and, above all, in large numbers. They come ashore in great splattering waves, little bodies bouncing and scrabbling *en masse* through thundering surf and slippery kelp fringes, dagger beaks and clawing toenails finding purchase as they scale near-vertical rock faces, hopping from ledge to ledge. For reasons known only to themselves, they invariably pick the most weather-beaten, wave-pounded windward shores of the islands they nest on, preferentially selecting those coasts bordered by steep cliffs, some of them 50 m (165 ft) high or more. Defying gravity, and undeterred by occasional cart-wheeling falls, they demonstrate expert rock-climbing and balancing skills, using big bouncing hops to secure well-known footholds. Everywhere around the subantarctic islands, the passage of millions upon millions of their little stubby feet — rudders underwater and cleated boots on land — have left deep grooves on impossibly steep rock faces, the Rockhopper's unfailing signature. But the chilling reality is that on more and more of these cliffs, that signature is all that remains of the vast throngs of past decades. For reasons yet to be fully explained and understood, the screeches and squabbles of their extinct millions have given way to the unbroken, mournful howl of wind and wave.

ABOVE The extraordinary head plumes of the Northern rockhopper are the most extravagant of any penguin, Gough Island, South Atlantic. OPPOSITE On New Zealand's subantarctic Auckland Island, a small colony of Southern rockhoppers (eastern race) enjoys a cooling bath under a waterfall.

SOUTHERN ROCKHOPPER

RIGHT Looking down some 75 m (246 ft) from the high volcanic cliffs of Monument Harbour, an adult Southern rockhopper completes the moult undisturbed by sea lion activity on the lower slopes, Campbell Island, New Zealand Subantarctic. BELOW Too late, an adult chases an egg-robbing skua, Sea Lion Island, Falkland Islands.

Southern rockhopper

Twenty-seven years separate two of my most delightful penguin experiences: two long and splendidly wild summers hiking and photographing in the Falkland Islands, the first in 1986 and again just before this book went to press. Never could I tire of watching their hyperactive, hell-for-leather lifestyles.

On my first visit, I happened to be camped by the largest colony in East Falkland, at an imposing headland called McBryde's Head, when the biggest storm of the season hit. Swaddled in my waterproofs and huddled behind large boulders, I watched in disbelief as rafts of returning penguins gathered offshore, preparing to land. Circling and porpoising excitedly for a while, as if mustering up the courage for the final approach, suddenly the entire group dashed for shore. They erupted from the roiling sea like little missiles hurtling toward the steep rock slabs above the kelp line. A few found purchase and immediately bounded up the face in a frantic bid to

outpace the next oncoming breaker. Only rarely did they succeed, the majority being washed helter-skelter back down to try again, and again — and again. Yet the steady flow of arrivals at the colony, dripping wet and crests dishevelled, were proof that their perseverance paid off.

Though I didn't know it yet, I was lucky to see such thronging masses of 'Rockies', as the Falkland Islanders call them. One sprawling colony I saw — intermingled with albatross nests on Beauchêne Island — was some 300 000 pairs strong. But by the end of that season something very strange had started to happen. It became evident on Pebble Island, then on Sea Lion Island, and even more so on Kidney Island. Soon after the chicks had fledged, the adult penguins returning to moult began dying en masse. In some places, dead bodies were piling up, providing a feast for scavenging Giant petrels and Turkey vultures.

It took years of research, tissue sample analysis, population monitoring and nesting success studies to

ABOVE Rockies, as they are nicknamed in the Falkland Islands, are inveterately curious, compelled to investigate any newcomer near the colony, Sea Lion Island. LEFT Strident braying marks a courtship greeting display, Deaths Head, West Falkland.

SOUTHERN ROCKHOPPER
ABOVE A well-timed landing involves staying just ahead of the next crashing wave, hopping frantically from ledge to ledge to avoid being washed back in.
RIGHT Exploding out of the water like torpedoes helps them gain a foothold above the slippery kelp line.
OPPOSITE Colonies are invariably sited high on the steep, storm-lashed windward sides of islands, where long queues wend their way up well-worn routes.
All West Point
Island, Falklands.

ascertain that the prime cause of death was simply a shortage of food; after the demands of the breeding season, adults could not regain sufficient condition to survive the three weeks of fasting that their mandatory annual moult requires, a time they are confined to land.

Starvation problems continued through the turn of the millennium, with poor breeding success and mostly shrinking colony sizes. Despite being studied at length, the causes remain unclear, though probably connected to a 2°C rise in sea temperatures. In 2002/3 a deadly red tide, or toxic algal bloom, dealt them a further blow, with numbers fluctuating since then.

A quarter-century after my first magical rockhopper encounter, I am delighted to find them finally doing well again. Although their colonies today are still mere shadows of their former glory, at last it appears they are enjoying a few good years. Plump and more hyperactive than ever, many parents are raising twin chicks, rotund and glowing with good health. Sea spray stinging my eyes, I watch their homeward throngs,

SOUTHERN ROCKHOPPER

RIGHT AND BELOW
Some colonies in the
Falkland Islands (Saunders
Island, top) have recovered
spectacularly in recent years,
while others (Sea Lion Island,
bottom) remain shadows of
their former selves.

BELOW RIGHT AND
OPPOSITE Unusual among
penguins, almost all colonies
are located near a source
of clean fresh water —
either a small spring or
stream — where guarding
adults can drink and bathe,
West Point and Saunders
Islands, Falklands.

SOUTHERN ROCKHOPPER

ABOVE Returning parents greet each other and their chick loudly for several minutes, Deaths Head, West Falkland.

RIGHT (BOTH) Even on flat ground, hopping is the favoured means of getting around on land; head and flipper posturing announces a new arrival, still wet from the sea, Sea Lion Island, Falklands.

many thousands strong after a good day's fishing, snaking their way in little hops up steep rocky ledges and over tussock-covered bluffs. Once again I feel sheer elation, the more so knowing how badly they had suffered.

In the intervening years, I visited other far-flung breeding islands right around the Subantarctic, as far as Macquarie Island on the opposite side of Antarctica, south of Australia. Many of them were even worse off than in the Falklands.

At Campbell Island's Penguin Bay, due south of New Zealand, I witnessed what a population decline of about 95% means for such a gregarious species. The small remnant patches of what had once been an enormous sprawling colony were surrounded and continuously harassed by disproportionate numbers of skuas, regularly swooping down to steal eggs and chicks.

Worse still were the Giant petrels who lined the access ways to the colony, plucking away and dismembering healthy but helpless adults returning from the sea in groups too small to defend themselves effectively.

Slightly farther north, the Auckland Islands were long thought to represent a healthy reservoir for the species. Yet all of the colonies I examined in 2004 were down to probably less than 100 nests each. Even the rockhoppers' behaviour here was totally different from what I'd known at larger breeding sites. Instead of nesting in dense aggregations out in the open, here they concealed themselves beneath boulders and thick clumps of vegetation, the adults arriving surreptitiously through narrow passages between rocks and quickly moving out of sight.

Only on Chile's Diego Ramirez Island, by far the species' southernmost colony at 56°36'S latitude, have Southern rockhopper numbers apparently not plummeted. This, and their tentative gains in the Falklands, provide much-needed hope for a severely embattled species.

ABOVE LEFT With a superb sense of balance, everyday commuters ascend high cliffs by hopping from ledge to ledge, where generations of traffic have grooved the solid rock faces, New Island, Falklands.
ABOVE RIGHT TOP Leaping feet-first off cliff ledges, while using wings for balance, Deaths Head, West Falkland.
ABOVE Once underwater, feet become rudders while flippers provide propulsion, New Island, Falklands.

Northern rockhopper

If the Southern rockhopper derives a charismatic character from its cocky headgear — fine curving feathers worn in a spiky diadem — it is totally and utterly outdone by its northern relative. The world's newest and most overlooked penguin species, the Northern rockhopper sports an extraordinary crown of flowing plumes, giving it a sort of panache that must be seen to be believed.

In 2006, an almost impossible dream came true the day I first sighted the distant outline of Tristan da Cunha Island on the horizon, at the end of seven days' sailing aboard a crayfishing vessel out of Cape Town. This imposing semi-dormant volcano, lost in mid-ocean about halfway between South Africa and Argentina, is the remotest inhabited island in the world. Here, 261 islanders, all British subjects, share the seven surnames of the founding fathers who settled on these daunting shores nearly two centuries ago, beginning with William Glass in 1816. I felt exceptionally privileged to be invited

by the Island Council to come here to photograph the island's remarkable natural wonders, particularly its albatrosses and penguins. Between them, Tristan da Cunha and its satellite islands are home to the entire Atlantic Ocean population of Northern rockhopper penguins; 80% of the world total for this poorly understood endangered species. The remainder is shared between two islands in the Indian Ocean, Amsterdam and Saint Paul, in the French Southern Territories.

Although the sea was windless the day I landed at the picturesque little settlement, known officially as Edinburgh of the Seven Seas, that soon changed. At just 37° S, a latitude roughly equivalent to Gibraltar or San Francisco in the north, Tristan's climate is technically temperate, but in actual fact it is remarkably subantarctic in nature. In addition, the massive shape of the volcano, which rises steeply to 2060 m (6758 ft), generates its own cloud cap, with swirling mists and changing winds an almost constant feature. A week passed after my arrival before the weather was again settled enough to

NORTHERN ROCKHOPPER
ABOVE With sturdy bill, red eyes and shaggy hairdo, a parent standing guard over its young chick puts on an intimidating look when fending off tresspassers, Gough Island.
OPPOSITE With young chicks to feed, long lines of parents stream down muddy pathways through the dense tussock grass, heading out on their daily fishing trip even before the first light of dawn, Nightingale Island.

NORTHERN ROCKHOPPER

ABOVE In the final stages of incubation, an adult crouches over its nest, the eggs held against a warm brood patch under a fold of belly skin.

LEFT (BOTH) Both greetings and courtship involve sky-pointing while shaking heads, with shrill braying and flipper waving.

OPPOSITE The Glen is one of many spectacular valleys where the species nests on steep bouldery slopes. All photos Gough Island.

NORTHERN ROCKHOPPER

ABOVE A day-old Northern rockhopper chick begs for food while both parents stand guard, Gough Island, South Atlantic.

make the crossing to famed Nightingale Island in the small police speedboat, guided by James Glass, then head of the Conservation Department.

On this picturesque little island, free of rodents and other introduced pests, nest about three-quarters of Tristan's entire population of rockhoppers. A veritable flood of penguins met me as I stepped ashore, hip-hopping in constant two-way traffic up and down the steep rocky landing. Although I'm not in the habit of comparing animals to people, the parallel here became overwhelming, the busy commuters' amazing headdresses morphing from regal tiara one moment to Rastafarian dreadlocks the next.

I followed busy penguin gangs to the interior of the island, an easy walk up 'the road', a swath cut by the islanders through the dense thicket of head-high tussock grass that covers much of the island. Unlike typical penguin islands farther south, here tussock grass grows as a continuous interwoven tangle of upright reeds, more like a miniature bamboo forest and nearly

as impenetrable. Columns of penguins followed the manmade pathway for a few hundred metres before ducking into well-trodden galleries through this tussock for the rest of the trek.

Guided by the sounds of squawks and squabbles, along with the soft peeping of newly hatched chicks, I followed on hands and knees to find their nesting area sheltered from both wind and rapacious skuas beneath the reedy canopy. Sooner than I imagined, I found myself in their midst, tucked away in a little underworld of straw. And all the while, strings of commuters padded by past my boots, while staunch pairs stood squarely in defence of their tiny bundles of black fluff.

It is heart-rending to think that a few short years after my visit, this tiny island would become the scene of a horrendous oil spill, killing untold thousands of this deeply beleaguered species whose worldwide population, like that of its southern relative, has shrunk by over 90% in historic times.

ABOVE Their breasts soiled from time in the colony, a group of Northern rockhoppers head down to the sea on Nightingale Island, site of the *Oliva* oil spill a few years later. Tristan da Cunha's iconic volcano stands out in the distance.

LEFT When subantarctic fur seals pass through the colony, penguins are often knocked off their nests, their eggs and chicks inadvertently crushed, Gough Island, South Atlantic.

NORTHERN ROCKHOPPER
Looking like a classic case of a 'bad hair day', wind buffets a beautiful young adult's crest this way and that, while squalls lash the rugged inland crags of Gough Island.

Two weeks after leaving Nightingale I met the Northern rockhoppers again in an even wilder setting, on neighbouring Gough Island, nearly 400 km (250 miles) to the southeast. There is no more fitting home for the world's most outlandishly dressed penguin: storm-lashed crags, pinnacles and cliffs; mist-filled canyons and wind-whipped waterfalls; verdant fern-fields and hanging bogs; and dripping forests of gnarly lichen-clad *Phylica* trees, dotted with stout tree-ferns.

Nesting on precipitous rocky slopes — where else would one expect a Rocky to choose — and at the mouths of glowing-green, gut-like valleys, they must elbow for their space among throngs of elegant but aggressive subantarctic fur seals, as well as deal with an exceptionally high population of skuas. On several occasions I saw fighting bull fur seals knock brooding penguins clear off their nests, while skuas are always on the prowl for such opportunities to snatch a chick. But with classic Rockhopper tenacity, they staunchly fight off threats, whatever the size or intent of the

opponent. My lasting memories of Gough are sounds of crashing waves and howling wind, punctuated by the sing-song wails of fur seals and the sooty albatross' haunting cries echoing down the canyons — and of course, the dishevelled golden plumes of little cocky penguins flailing this way and that with every wild gust of salty wind. 🐧

ABOVE AND LEFT Flowing golden crests were once used by crafts ladies on Tristan da Cunha to fashion delicate shawls, Nightingale Island.

5
Fairies of the Night: An Odd Couple

A painterly summer sunset slowly transforms the sky from limpid blue to intense gold, transforming splashing wavelets along the western shore of Phillip Island into glittering tinsel. We wait. The mantle of twilight envelops us. Wallabies emerge from the inland thickets to graze the coastal bluff in the coolness of night. Then we see them at last: the first raft of tiny penguin heads dancing fleetingly in shimmering coppery pools between waves. Even though they are barely visible, my friend and I — tripods and cameras at the ready — begin shooting. We both feel these photos will be far more telling than any in-your-face close-ups we might get at the nest.

This is the essence of the diminutive nocturnal species known officially as the Little penguin, the smallest of them all. In New Zealand it is known by the more descriptive name of Little blue penguin, whereas the Australians sometimes call it the Fairy penguin — the name I like best, as it captures the spirit of this elusive little character.

Phillip Island, on Australia's south coast near Melbourne, represents one of a handful of happy stories amid the otherwise sad saga of a species deeply affected by human activities. From the introduction of feral predators, to the destruction of coastal nesting habitat and depletion of food stock by fisheries, the numbers and range of this species have followed downward trends for many decades. But on this island, in a private programme largely self-funded through well-managed penguin-viewing tourism, both human residents and penguins benefit. At the same time, this project also supports one of the most comprehensive and long-term research programmes on any penguin species.

As stars begin to pepper the darkening night sky, the pace quickens all along the rocky beach where we are busy photographing the arriving penguins. Flock after flock, some several dozen strong, tumble ashore as waves cascade over the slippery boulders. That's when I appreciate just how small they really are, standing

ABOVE Peering from the wavewash, a timid Little blue returns to feed its chicks in the dim glow of dusk, Phillip Island, Southern Australia.
OPPOSITE Shy and mostly solitary, a Yellow-eyed penguin makes its way through tussock grass toward the interior of Enderby Island, where its nest is hidden under thick vegetation, New Zealand Subantarctic.

LITTLE BLUE PENGUIN
ABOVE As the last colours of sunset fade in the sky, returning flocks gather in growing numbers along a secluded stretch of shoreline before making a dash for their nests inland on Phillip Island, where a remarkable conservation success story is fuelled by tourist dollars, Australia. RIGHT Chicks await the return of their parents at the nesting burrow entrance, Bruny Island, Tasmania.

barely 30 cm (12 in) high and 1 kg (2¼ lb) in weight, only slightly larger than a 1 L (1 quart) milk bottle. They shake and preen themselves dry, their numbers swelling as more return from the sea. Then ripples begin to form among their ranks, as the first arrivals peel away to scurry up the steep slope in tight bunches. Unlike larger penguins, they are nervous and utterly quiet, quickly moving up well-trodden, sinuous pathways winding through the coastal vegetation. Yet, wary as they are, if I anticipate their move and keep quite still, they walk right on past with no more than a glance in my direction. Some even walk directly over my camera, which I'm firing remotely on the ground. When they reach flat ground 20 m (65 ft) or so above sea level, they fan out, melting away into the shrubby groundcover where their nest burrows are secreted. Only the discreet peeping of hungry chicks reveals their well-hidden activities.

Within days I find myself on a different shoreline, but this time I'm looking east toward the first hint of dawn. I've just spent all night with the Fairies and, though

I haven't slept a wink, I'm too enthralled to be tired. This is Bruny Island off the eastern shore of Tasmania, where The Neck Nature Reserve (a narrow stretch of sand dunes representing the midriff of the island) is a favourite penguin and petrel nesting site.

When I arrived in the late afternoon, I wasn't prepared for the busy nightlife that would take over under cover of darkness. A handful of gulls and a pair of crows lazily patrolled the long crescent beach, where a faint line of little footprints in the fine windblown sand barely hinted at penguins passing. Well after the last sunset glow had faded, first arrivals appeared surreptitiously in twos and threes. Surprised by the unusual object near their pathway, they pecked at my glinting camera lens inquisitively before vanishing into the scrub.

By midnight all penguin traffic had ceased, but now the dunes resonated with strange peepings and mewings, and plaintive brays uncannily resembling a human infant crying gently. Concentrating on the

ABOVE Using well-trodden pathways up into the dunes of Bruny Island, small groups shuffle across the sand, unsuspecting of the remotely triggered camera.
LEFT Crouching in the rank grass on the road verge, a nesting bird prepares to dash across a coastal road, a risky move that causes substantial mortality throughout much of the species range, Bruny Island, Tasmania.

sounds, in the faint starlight I could barely make out pairs courting and mating amongst the bracken and beach grass, or scuffling about scratching burrows and carrying tufts of grass to line their underground nests. Every once in a while a fountain of white sand came flying up from where someone was doing their house cleaning. The more mature breeders all had healthy pairs of fat chicks who, emboldened by the intense darkness of this moonless night, emerged from their burrows to stretch and explore.

Sometime after midnight came a slight lull, with families dozing quietly together outside their burrows. Activity resumed in the small hours, as flocks of Short-tailed shearwaters began to arrive. Swooping and circling overhead on the freshening sea breeze, they added their bizarre guttural calls to the eerie nocturnal concert. Soon they were crash-landing among the penguins and scurrying about in search of their own burrows.

I barely noticed the first morning glow appearing

between the gathering clouds, but for the night Fairies — penguins and shearwaters alike — this signal was as clear as an alarm bell. In a sudden scramble, the chicks retired deep in their burrows and everyone else quickly disappeared back out to sea. As the soft twilight hues caressed the now-deserted beach, silence pervaded once more. I gazed dizzily at the intensifying colours on the horizon, where I knew the tiny penguins were swimming toward their feeding shoals — oh, what a night!

White-flippered penguin

The Little Blue penguins in New Zealand are the same species as Australia's Fairies, but here they are distributed in many smaller colonies sprinkled around the coastlines of North, South and Stewart Islands. But New Zealand offers another intriguing twist to the Little penguin story: a small population nesting only around Banks Peninsula, along the east coast of

LITTLE BLUE PENGUIN
ABOVE Emerging from a freshly dug nest burrow, Bruny Island, Tasmania. LEFT During the middle of the night, parents and chicks relax outside their nest burrow at The Neck Reserve, Bruny Island, Tasmania. OPPOSITE (ALL) Surreptitious comings and goings take place in the half light of evening under faint coppery hues on Phillip Island, while sunrise on Bruny Island reveals multitudinous footprints of nocturnal commuters.

ABOVE Instead of using burrows some open nests are located inside totally lightless caves, such as this incubating bird near Charleston, West Coast, New Zealand.

RIGHT (BOTH) At the nesting colony, everything takes place under the cover of darkness, with commuters arriving in the evening (top, Phillip Island, Australia) and departing again at the crack of dawn (bottom, Bruny Island, Tasmania).

LITTLE BLUE PENGUIN

ABOVE Navigating through coastal thickets can be a daunting exercise for such a tiny penguin, Bruny Island.

RIGHT Large chicks are tended by their parents just outside their nest burrow in the early evening, Bruny Island, Tasmania.

BOTH PAGES The White-flippered subspecies is restricted to Banks Peninsula along the east coast of New Zealand's South island. By far the largest mainland nesting colony is found along the shores bordering the Pohatu Marine Reserve, where a penguin-friendly farm, Pohatu Penguins, provides a safe haven, with feral predator trapping, vegetation replanting and nest boxes provided, partly funded by cautious penguin-watching visitors.

the South Island, is so different from all others that it is sometimes regarded as a separate species, the White-flippered penguin. Sadly, however, whether species or subspecies, the rate of attrition for this population is as bleak as most others — with one exception. On the wild outer coast of the ancient volcano forming Banks Peninsula is a deep inlet where two people have become the penguins' best friends.

Leaving earthquake-wracked Christchurch city, I drive out over the craggy rim of the long-sleeping volcano, and on along sinuous ridges where sheep and cattle graze in golden summer pasture overlooking deep cobalt blue bays and inlets. Ice Age glaciers once carved narrow gouges through basaltic lava layers, sculpting the old volcano into the shape of a gigantic water lily flower resting on the sea. The largest of these inlets makes up Lyttelton Harbour, another shelters Akaroa, an early French settlement in New Zealand. All used to be prime White-flippered penguin nesting habitat. Hammered by introduced predators, both domestic and

feral, one colony after another has shrunk and vanished, a process regrettably still ongoing.

I drive on. Two hours out I crest one final ridge, and without warning the gravel road plummets toward the open ocean — there is no land between here and South America. At the bottom of the track, tucked at the back of a spectacular miniature fiord called Flea Bay, nestles a small farmhouse and ample sheep-shearing yards. I've reached 'Pohatu Penguins', the private initiative of a remarkable farming couple, bordering the Department of Conservation's Pohatu Marine Reserve.

Since 1991, Shireen and Francis Helps have made it their mission to save the penguins of this bay. Funded by revenue from a small flow of penguin-viewing tourists, they relentlessly trap ferrets and stoats and other introduced predators, such as feral cats. They've also replanted much coastal vegetation, sprinkled their land with safe penguin nest boxes, and even sometimes fed starving chicks to boost survival in bad years. The result is the largest mainland penguin colony

YELLOW-EYED PENGUIN
RIGHT An adult looks out from its nest on Enderby Island, part of the Auckland Islands group. OPPOSITE (TOP) Well hidden in *Dracophyllum* and *Blechnum* fern thickets of Northwest Bay, a parent relaxes with a nearly fully grown chick on Campbell Island, New Zealand Subantarctic.

— of any species — in all of New Zealand. With over 700 active nests most seasons, this is an extraordinary achievement and an inspiration for what individual perseverance, applied with passion and in the right place, can achieve.

Yellow-eyed penguin

New Zealand is home to not one night-loving penguin, but two. The rare Yellow-eyed penguin is not as strictly nocturnal as the Little blue, but is certainly one of the shyest. It usually commutes to and from its isolated nests in dense vegetation at dusk and dawn. This is an enigmatic species that some believe to be the oldest living penguin in their evolutionary lineage, having separated from other living species possibly as long as 15 million years ago. Adding to the mystery is the recent discovery of a very close, now-extinct relative that preceded it on New Zealand's main shores up until about 1000 years ago. Similar in shape but slightly smaller, the Waitaha penguin apparently was

exterminated when the first humans arrived from the Pacific islands, and was replaced by the Yellow-eyed, who colonised from southerly subantarctic islands. Why one species went extinct while another was able to establish itself in its place is a mystery that will no doubt endure.

Today the Yellow-eyed penguin's main nesting areas in New Zealand extend from Foveaux Strait north to the Otago Peninsula, where another penguin-friendly farming family offers them safe haven in a private reserve, providing them will specially made nesting shelters. At 'Penguin Place' eager visitors are offered daily tours through gallery walkways hidden in the sand dunes, from where they can spy on the shy penguins without alarming them. The revenue serves to finance predator control and revegetation projects to improve their habitat.

Far to the south, subantarctic Campbell Island is the Yellow-eyed penguin's native stronghold. Well within the latitudes known as the Furious Fifties — between

LEFT Golden eyes and streaked yellow feathers around its face give this species a unique appearance, Enderby Island.

FAR LEFT Protected from rain in a well-used nest under a thatch of dead fern fronds, a parent stands guard over its well-camouflaged sleeping chick, Northwest Bay, Campbell Island.

BELOW A nest tucked under a canopy of ferns and nettles amid wind-carved coastal scrub, all photos Enderby Island.

YELLOW-EYED PENGUIN

ABOVE At their nest on Enderby Island, this pair's mutual greeting consists of repeated deafening screams that gave rise to the species' Maori name of 'Hoiho', meaning 'noise-shouter'.

LEFT In the dark sanctity of the undergrowth, a chick is about to be fed.

RIGHT Greetings and courtship are often marked by jerky head movements and silent, rigid posturing.

FOLLOWING PAGES Unseen from the stormy outside world, commuters move quietly across the mossy ground beneath the intertwined canopy of dense *Dracophyllum* on Campbell Island, the species stronghold on this southernmost of New Zealand's subantarctic nature reserve islands.

YELLOW-EYED PENGUIN

ABOVE Infrequently, small groups may get together to head inland when returning from the sea, seen here wending their way through thick tussock grass on Enderby Island.

RIGHT Pair greeting.

FAR RIGHT To cool off on a sunny, windless day, barely feathered flipper undersides turn bright pink as blood is flushed to the skin, both Enderby Island.

the Roaring Forties and Screaming Sixties — the island is strafed by rain over 300 days a year, and winds above 40 kph (25 mph) blow at least six days a week. The maximum wind speed was recorded gusting on a ridge top at an incredible 225 kph (140 mph). Nearing the end of our four-month private expedition, our little 13-m (43-ft) sailboat *Mahalia* lay snug on her two anchors, untroubled by the raging storm funnelling down the long arm of Northeast Harbour, once the hang-out of whalers and sealers. As usual, the wind was screaming through our rigging, at times tearing sheets of water from the dark sea surface and mixing it with horizontal rain. Yet to my amazement, high-pitched screams could be clearly heard rising above the roar of the tempest from beyond the shoreline 100 m (330 ft) away.

The loudest of all penguins, a pair of Yellow-eyeds were courting somewhere on the dark scrub-clad slope. In gathering twilight, I followed their sound to shore by dinghy, then ducked into the dense, almost impenetrable *Dracophyllum* thickets, this so-called dragon-bush being the only woody 'tree' growing on the island — all of 3 m (10 ft) tall. Sheltered from both wind and light beneath the tightly knit, dripping canopy, sporadic calls led me onward. I crawled and twisted between interwoven branches, silently moving deep into a strangely quiet, soggy underworld of gently gurgling rivulets, punctuated by the unnerving growls of unseen Hooker's sea lions lurking in the gloom. Tantalisingly, the penguin calls lured me onward but never seemed to get any closer. Eventually I was thwarted by the oncoming night. Only later, at Northwest Bay on the opposite side of the island, did I find the penguins en famille — shy little creatures in a dark green elfin world of springy basket-ferns, spongy moss cushions and frilly lichens, a fitting home for one of the world's rarest, most secretive penguins.

ABOVE Most often commuters travel alone, like this one among *Anesotome* megaherbs on Campbell Island, sometimes walking several kilometres to their nest.

6
Monarchs of the Far South: Kings and Emperors

Of today's 18 penguin species, it is the true royalty — the Kings and Emperors — that most resemble the giants that ruled the seas of the southern hemisphere many millions of years ago, long before increasingly adept mammalian competitors and predators gradually usurped their space. But in their own ways, today's lesser giants are still the rulers of the far south. One we know as a creature of the ice, the other as occupying the wildest expanses of the Southern Ocean and its widely scattered subantarctic islands. Yet, this conception is derived from what we see of their breeding habits. Satellite tracking studies indicate that during the long winter months they may actually be rubbing flippers on the same feeding grounds along the edge of the frozen Antarctic seas. Indeed, what distinguishes the two species most markedly is the way in which they have resolved their shared problem: how to raise chicks that, because of their large size, cannot grow to independence within one short Antarctic summer. Their solutions

ABOVE Emperor penguins at home on the frozen sea near the Ekström Ice Shelf, Weddell Sea.
LEFT A pair of King penguins courting, Macquarie Island, Australian Subantarctic.
OPPOSITE King chicks were once referred to as 'Oakum Boys' by sailors because of their shaggy coats. Here, a large chick has found its own private island in a meltwater stream away from the crowds on South Georgia, South Atlantic.

KING PENGUIN
ABOVE Just a few days old, a chick emerges to be fed from under its parent's protective feathered blanket.
ABOVE RIGHT TOP Appearing larger than its parent because of its ample down layer, a chick has an insatiable appetite during its final growth spurt before fledging, Gold Harbour, South Georgia.
ABOVE RIGHT BOTTOM With lightning speed, two males flail their flippers during group courtship, while the female looks on, Volunteer Beach, East Falkland.

are as different as they are extraordinary. The Emperor heads south, setting out on a nine-month breeding season that begins in the dead of winter, whereas the King goes north toward the lush green islands of the subantarctic to begin its protracted cycle, disregarding all seasons and taking up to 18 months to raise a single chick. Additionally, both Kings and Emperors share the extraordinary habit, unlike almost all other birds, of not building a nest but carrying their eggs about with them, balanced atop their feet and tucked under an ample blanket of loose, feathered, abdominal skin — Emperors shuffling on the smooth frozen sea; Kings selecting flat outwash plains below canyons and glaciers.

King penguin

There are few more awe-inspiring sights than looking down from the vantage of a tussock grass ridge, upon something approaching 100 000 King penguins and their chicks. The colony spreads out in a living carpet moulded to the contours of the land. Seen

from a distance, graceful veins of cinnamon brown create swirling patterns on the plain, like freeform marbling on a foamy cappuccino, where shaggy hordes of rotund pre-fledging chicks gather along cool meltwater streams. The melodious two-tone trumpeting calls of the adults mingle with the incessant warblings of a sea of chicks, waves of sound rising directionless whenever the fierce, sleet-laden gusts abate. These mind-bending scenes, with some variations, are repeated on just a few islands sparsely dotted near the Antarctic Convergence, or Polar Front, a fluctuating line wavering mostly between 50° and 60° S, where frigid Antarctic waters slide beneath the more temperate surface waters of the Southern Ocean.

I have marvelled at such a King penguin megacolony on South Georgia in the South Atlantic, where katabatic winds tear through heavily glaciated valleys amid sawtooth ranges, with Light-mantled albatrosses circling overhead and elephant seals

ABOVE AND BELOW Group courtship and mating sequence, Volunteer Beach, East Falkland.

KING PENGUIN
ABOVE RIGHT Strolling down the beach under an afternoon rainbow.
RIGHT (BOTH) A flipper flapping argument erupts between two males bathing in a shallow lagoon.
BELOW Male advertising for a mate by sky-pointing with dual-note trumpeting. All photos Volunteer Beach, East Falkland.

snorting along dark sandy beaches. But it is the setting of Macquarie Island, below Australia and almost directly on the opposite longitude to South Georgia, that impressed me most — not least because of its poignant history.

Approaching the verdant palisades of Macquarie, the sea appears to boil as dense hordes of curious Kings swim out to meet us, leaping and splashing in a tight melee. Only as we draw nearer to land do I become aware of the jam-packed masses also crowding the coastal shelf of Lusitania Bay. That's when I realise that the rafts of penguins surrounding our ship represent only the advance guard. By the time I'm driving my inflatable boat toward the beach, this escort is so fired up that many of them jump clear out of the water to take a look inside the boat, occasionally bouncing ungracefully off the sides of the rubber pontoons. As I cruise along the shore, the din of the breeding colony wafts across the beach with a kind of pulsating rhythm, but the strange sight of rusting machinery towering in their midst speaks of a dark past.

In 1889, a man named Joseph Hatch pulled up to this same beach with a very different intent to mine. Whales everywhere were becoming scarce and the world hungered for oil to fuel booming industries in the northern hemisphere. His attention first focused on the writhing herds of elephant seals lining the beach, which were slaughtered and rendered into oil. But having soon finished off these blubbery giants, his gaze next turned to the penguins. Over the next four decades his enterprise boiled down an estimated three million penguins, using the desiccated carcasses coming out of the cauldrons to fire the boilers. Dead penguins fuelled perpetual flames as oil barrels lined the beaches of Macquarie until there were just 3400 Kings remaining, a mere fraction of a percent of the initial population. Strictly protected ever since, their numbers are still recovering today, taking almost a century to approach half a million.

Similar stories took place elsewhere, as greedy men plundered the biological riches of the Southern Ocean.

LEFT Great rafts of bathing King penguins gather offshore from their largest colony at Lusitania Bay, Macquarie Island.

FOLLOWING PAGES A common sight on many busy beaches of South Georgia, commuting King penguins negotiate their way between groups of breeding elephant seals at Gold Harbour.

KING PENGUIN

ABOVE Walking slowly, a group returns to the sea through a blinding sandstorm.

RIGHT On a rare, blistering hot, windless day, moulting adults seek relief by a mirror-smooth rain pond. Both photos Volunteer Beach, East Falkland.

FAR RIGHT A departing group swims past sparring young elephant seal bulls at Macquarie Island.

King penguins vanished from the Falkland Islands, but have likewise been making a valiant comeback with small colonies now at several scattered locations. The resurgence of the species has granted it the conservation category of Least Concern, but in actual fact new threats may yet lurk on the horizon, dangers generated by human activity that affect all penguins no matter what their lifestyle. These range from over-fishing to ocean pollution and global warming.

In the King penguin's case, rising ocean temperatures are causing the slow southward shift of the average position of the Polar Front, the rich oceanic zone where their prime food consists almost exclusively of lantern fish and other deep-sea species caught at depths reaching 440 m (1445 ft). With the Polar Front displacement taking place at around 40 km (25 miles) per decade, and no other available nesting islands in these southern latitudes, the commuting distances will eventually become too great — maximum round-trips are about 700 km (435 miles) — for parents to sustain

their young. One study at the Crozet Islands, where the next largest colonies can be found, shows that such cataclysmic times for this King penguin population could already be less than four decades away.

BELOW Not normally a penguin of snow and ice, adult Kings struggle to keep their balance after a spring snowfall at Grytviken, South Georgia.

BOTTOM Landing in surf can be tricky for such heavy penguins, Salisbury Plain, South Georgia.

KING PENGUINS
RIGHT A courting pair performs its stretching display, their heads rising ever so slowly in unison until freezing at full height for half a minute or more, Volunteer Beach, East Falkland.

CLOCKWISE FROM ABOVE The characteristics of colonies change with the seasons. In spring, moulting non-breeders segregate from large chicks that made it through the winter at Salisbury Plain, South Georgia; early summer at Volunteer Beach, most birds are sitting on eggs; numbers swell when chicks begin to hatch and more birds are courting in late summer; getting ready to fledge, last year's chicks will soon exchange down for feathers, South Georgia.

Emperor penguin

The Australian icebreaker *Aurora Australis* is still ramming its way metre by painstaking metre into the solid fast-ice, hard as concrete, as we approach the shores of East Antarctica, heading toward Davis Station. A few simple words ring out, the ones I've been waiting to hear with barely containable anticipation: 'Get ready, we have the perfect weather window!' Sponsored by the Australian Antarctic Division, I am here to accompany renowned Emperor penguin researcher Barbara Wienecke to her study site at the Amanda Bay rookery, near the Amery Iceshelf in Prydz Bay. This place is about as far removed from our so-called 'real world' as I could ever hope to reach.

We are jolted into action after 16 days at sea. Gear is assembled, the helicopters readied and, before we know it, we're airborne — two small bubbles buzzing across the vast blue sky of the Antarctic. Below us, the world changes scale. The tidy cluster of colourful buildings clinging to bare rock in the Vestfold Hills — a

fulcrum of Australian Antarctic science since 1957 — look like mere matchboxes in the immensity of their surroundings. Likewise our bright red icebreaker shrinks to toy-like proportions, dwarfed amid giant glistening icebergs the size of city blocks.

We fly high to avoid disturbing nesting petrels and Adélie penguins. Below us pass a kaleidoscope of striated rock formations, crystalline frozen lakes, crenellated glaciers, turquoise ice-cliffs and aquamarine pools of open sea water, where Snow petrels flutter like confetti. To the south stretches the domed infinity of the Polar Icecap. Northward is a blinding mirror of sea-ice reflecting the spring sunshine. Some 50 km (31 miles) out we draw a slow circle over a large embayment in the ice shelf, then begin angling down toward a chocolate-brown rocky island at its centre. That's when I first pick out the colony; by the telltale greenish-brown guano smudge on the otherwise pristine sea-ice nearly a kilometre beneath our rotors. For an

EMPEROR PENGUIN

RIGHT TOP During the last days of December, as the sea ice near the nesting colony is breaking up, chicks eventually build up the courage to enter open water for the first time, even though still partly covered in down and only about half their eventual adult weight. Flailing awkwardly, their first task is to learn to swim, Cape Darnley, Davis Sea, East Antarctica.

RIGHT By the time it reaches adulthood, an Emperor is the most accomplished diver of all penguins, capable of remaining submerged for long periods and reaching amazing depths (22 minutes and 564 m have been recorded).

BELOW Details of exquisite adult plumage.

BELOW RIGHT Parent and chick share a tender moment, Amanda Bay, East Antarctica.

instant, a flashback transports me almost a quarter of the way around Antarctica where, as expedition leader on a Russian icebreaker 15 years ago, I took another helicopter flight to find a 'lost' Emperor penguin colony that had last been sighted three decades previously. The Lazarev Iceshelf had completely changed shape since it was last surveyed, so we had no clues where to search. Long after midnight, with the sun hovering near the South Pole, we finally spotted the penguin cluster, like an eerie apparition between golden icebergs more than 10 km (6 miles) from their original location.

Now I once again feel a tingle of awe run through me... How can any warm-blooded creature choose to live the way the Emperor penguin does, ensconced in a landscape that looks as devoid of life as the face of the moon? The Emperor's life story has been recounted in documentary films and books on Antarctica, yet it remains so extraordinary that it is worth telling anew.

At the onset of the long Antarctic winter, when sunshine retreats and the sea freezes over, most

warm-blooded animals start heading north, but not so the Emperors. Feeding near the periphery of the ice in summer, they alone begin heading south with the winter night, first swimming, then walking for tens if not hundreds of kilometres to their traditional breeding spots, usually sited on fast-ice sheltered from the wind by icebergs or glaciers. As the sun dips below the horizon for the last time in May, the females lay their single huge egg, which they quickly pass over to their mates and then leave. The males will endure the next two months of darkness without food, carrying their precious cargo on top of their massive feet. By shuffling into tightly packed huddles to collectively survive the most intense cold experienced by any animal — down to minus 60°C — they constantly rotate places so that each bird spends relatively little time fully exposed to the elements. The females begin the long trudge back across the ice as the first colours of the returning sunlight grace the horizon, fat and sleek from their time at sea. With the spring sea-ice at its

ABOVE Expansive vistas of ice and sky, tinted by the midnight twilight of East Antarctica, presents an ethereal setting for the world's largest penguin, Amanda Bay. LEFT A chick's goggle-faced markings invariably evokes the 'Ahhh' response in humans.

EMPEROR PENGUIN
ABOVE (BOTH) As long as
air temperatures remain
below freezing, the
easiest form of long-
distance travel is by
gliding over the frozen
surface on their stomachs,
toboggan-style; but when
snow becomes sticky in
warm weather, they are
forced to walk, which is
more energy-costly,
Kloa Point.

RIGHT (BOTH) Massive
reptilian feet, reminiscent
of a dinosaur's, are needed
to negotiate a body
weighing up to 40 kg over
snow and ice, Prydz Bay.

maximum extent, their return journey may cover up to
200 km (124 miles) or more, but even so they often time
their arrival perfectly to take over the care of the freshly
emerged hatchlings. Astoundingly, even if they are late,
the males can still muster a few beakfuls of concentrated
food secreted from their own stomachs to keep their
new babies going for a few days. Chicks grow steadily
through the spring and early summer, ready to take to
the sea at about five months of age, weighing just 10
kg (22 lb) or so, only a third of their parents' weight. All
being well, their departure will coincide with the summer
break-up of the ice upon which they were raised.

When my thoughts jolt back to the present, our gear
is piled on the ground and the helicopters have departed,
leaving behind a silence so profound and so pure that it
seems a sacrilege to break it by speaking. Looking out
from the top of our frozen island, we are surrounded
by space so expansive I am compelled to spin around
several times to take it all in, breathing air so crisp and
clean it almost makes me dizzy.

Soon we've walked over a ridge and sat on a rocky
promontory quietly surveying the mesmerising scene
below: 10 000 Emperor penguins and their young
spread out in all directions. The windless air is alive
with the squeaky yodels of chicks, punctuated here and
there by the sensuous double-toned trumpeting of an
adult. Big groups of pearly-grey youngsters, already four
months old and excited to discover their own mobility,
set off exploring this way and that, sometimes locking
step with a passing adult or just egging each other
on toward one of the many subcolonies dotting the
icy landscape. They crane their necks in all directions,
intrigued by everything they encounter, their look of
surprise accentuated by their white-goggled faces.

The next morning is a hot one, the temperature
hovering only just below freezing. The chicks now take
turns scrambling up a slippery ice hummock, pecking
away at clumps of clean snow to keep themselves
cool and hydrated. Meanwhile throngs of non-breeding
adults, probably the younger ones prospecting for

EMPEROR PENGUIN
ABOVE A column of returning adults trudges resolutely along the face of the Flatnes Ice Tongue on their way to the Amanda Bay colony.
OPPOSITE TOP Tobogganing Emperors navigate unerringly across their vast feature-less world, Weddell Sea, West Antarctica.
OPPOSITE BOTTOM A rotund adult walks over wind-polished sea ice, where entrapped tabular icebergs catch the evening sun, Prydz Bay, East Antarctica.

future seasons, peel away from the colony to give us a close inspection. Sprawled on the snow enjoying a cup of hot tea, we find ourselves looking up as they tower over us, arching their heads down for a closer look. Their rugged, scaly feet look rough and heavy like those of a dinosaur, the snow crunching under 30–40 kg (66–88 lb) of penguin. Yet their white breast feathers are as shiny as silk, and their glowing orange 'ear patches' look as velvety as gold brocade — an incredible combination of rugged functionality and refined natural elegance.

From high on the island I can see lines of adults still returning from the invisible sea far to the north. Travelling beneath the magnificent face of a cobalt-blue glacier, some are tobogganing on their stomachs while others walk resolutely upright with a gentle, rhythmical sway of their necks. With the evening sun at my back skimming the icecap, I know that this is a special time in the Emperor's calendar. In about six or seven weeks, with the approach of midsummer's day, the arrivals

will stop, and the chicks will be left to make their own decisions as they make their way toward a new life at sea.

Once again my thoughts are transported to other times and places. The date was 19 December many years ago, and the place was Cape Darnley on the far side of Prydz Bay. That day, by sheer luck, I witnessed the departure of the Emperor chicks, freshly clad in that amazing waterproof coat of new feathers, the densest of any bird, that allows penguins to be who they are. The nesting colony was still about 3 km (nearly 2 miles) from the disintegrating ice-edge, but hordes of chicks, bereft of parental food supplies, were just beginning to leave, trekking across the ice toward where their parents had gone. At their first-ever sight of open water they seemed fearful and hesitant, gathering along the ice-edge in agitated bands. A stumble produced the first unplanned dive, followed immediately by a cataract of toppling bodies, splashing and squawking

EMPEROR PENGUIN
ABOVE Four-month-old chicks explore the vicinity of an arched tabular iceberg, Atka Bay, Weddell Sea.
OPPOSITE TOP Greeting between parents and chick.
OPPOSITE TOP FAR RIGHT From the air, a dark smudge reveals the Amanda Bay colony in early November, sea ice still stretching to the horizon. Discernible in satellite photos, such stains allow detailed counts all around Antarctica.

in confusion until their instinctive diving skills kicked in. As twilight fell and our ship pulled away, my last view was of a gaggle of grey-and-white chicks half feathered and half downy, the survivors of the rigorous winter that had already claimed many younger ones, sitting together on an ice-pan drifting out to sea, ready to start a whole new life.

The rarest of all gemstones in my penguin memories, when nothing could be better than the here-and-now, had taken place just three days before the Cape Darnley exodus. The colony at Kloa Point had not yet started to disband, with the ice-edge still far away and plenty of adults commuting back and forth. It was a perfect evening, the midnight sun skimming the glaciers and casting blood-red streaks of light onto the ice even as a distant gathering storm painted a blue-black sky for backdrop. I walked away across the sea-ice, drawn irresistibly in the wake of the penguin train, their long lines weaving sinuously between gigantic sunlit tabular icebergs, whose bases were grounded on the seafloor

hundreds of metres below the frozen sea surface. With no wind stirring the air, the silence was absolute. Then a soft, almost imperceptible sound caught my attention: the bell-like trills of courting Weddell seals, unseen in the depths beneath the ice, were resounding off the iceberg wall. Enthralled, I stopped and listened while watching the Emperors disappear into the distance. For a few indelible moments, I felt as if I had become part of another world — the fantastic world of penguins.

RIGHT From the Riiser–Larsen Ice Shelf to Amanda Bay, summer heat and winter blizzards take their toll on young chicks, but provide food for skuas during the spring thaw.

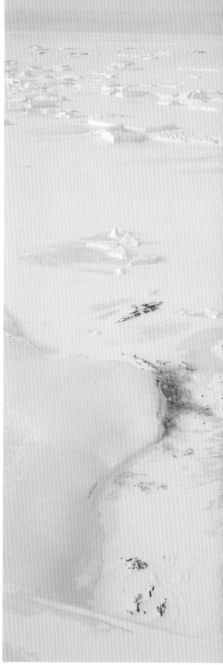

EMPEROR PENGUIN

RIGHT Soon to take to the sea for the coming three months, a pair near the Ekström Ice Shelf looks out at the wavelets, as stranded icebergs are liberated from the fast ice during the summer breakup.

BELOW Eerie stillness precedes a massive storm over Edward VIII Bay in East Antarctica, where a group of Emperors rests on a large ice floe.

ABOVE Fluffy chicks crowd the ice seemingly to the horizon off the Princess Martha Coast, Weddell Sea.

LEFT After a long search by helicopter, spotting a colony not seen for almost 30 years near the Lazarev Ice Shelf is a startling sight, the only sign of life in the tranquil vastness bathed in golden midnight sun.

2

SCIENCE AND CONSERVATION

Mark Jones

Penguins and People: A Retrospective
Mark Jones

I have often had the impression that, to penguins, man is just another penguin — different, less predictable, occasionally violent, but tolerable company when he sits still and minds his own business.
— Bernard Stonehouse, veteran Antarctic biologist, writer and educator, 1968

My worst penguin day

'Ohh, look mommy, look… that man's got a penguin hanging from his finger…'. I still cringe at the memory, high-ranking in my life's most embarrassing moments. I'd long prided myself as being sensible regarding the idiosyncrasies of nature, able to intuitively interpret an animal's behaviour, empathise with its character and truly respect and appreciate its world and its ways. Yet there I was, blood dripping from my hand, with — yes — a Magellanic penguin firmly clamped to my forefinger.

His flippers flailed excitedly as I repeatedly tried to pull away, dragging him from his swimming pool, his razor-sharp, slightly serrated bill sawing my flesh to the bone in a tug-o'-war (and of wills) I was clearly losing. He hung on: 'Being a tenacious little devil, aren't you?' I tried to act nonchalant while kiddies looked on, aghast, and mothers shied away, quickly ushering their dear ones clear of the sloshing fray.

'Darn… this hurts!' Desperate to disengage, I finally attempted to grab the irate bird by the scruff of the neck. Seeing this, at last he let go, splashing around in tight circles, wound up and expectant, looking to have another go at me. I quickly retreated to bind my lacerated finger... 'Hey, folks, no big deal!' I quipped to the hushed, shocked families, secretly nursing my humiliated ego.

Cute and cuddly, boys. Cute and cuddly!
— Skipper the Penguin (Tom McGrath), after escaping the Central Park Zoo, *Madagascar* (animation movie), 2005

Luckily, the deeper scars from how blindly dumb I'd been faded faster than my bemused remembrance of the afternoon I casually reached over the low-walled enclosure of the outdoor penguin pool at San Diego's SeaWorld. That day, instincts and well-honed field-skills that had stood me in good stead in all manner of wild environments had counted for naught. Lulled by laughing crowds and the distant hum of city traffic, snack bars and gift shops brimming with plush animal toys, I'd fallen for the stereotypic cute and cuddly image of penguins. But then, I wasn't actually 'out in nature'; those dapper little characters strutting and swimming in the California sunshine weren't 'wild'. I'd overlooked the fact that my action would be taken as trespassing his personal space.

Still, I'd learned my lesson: Never – ever! – hold out your hand to a 'Mag', as they are sometimes called for short. As many a penguin researcher will attest, the Magellanic can be one of the feistiest penguins, defending itself admirably with sharp, jabbing beak, scratching feet and brutal, bruising volleys of bony, beating flippers. This makes absolute sense when considering that, unlike many of their brethren found only on isolated oceanic islands, Magellanics are at home on the shores of continental South America, sharing their habitat with many wily predators. From the confines of deep nesting burrows, they deftly repel the likes of foxes and raptors, and even people. Essentially, they shouldn't be trifled with, and anyone who does encroach will probably think twice next time.

Educational experience

During my career as a wildlife photographer, I've been fortunate to spend many, many memorable hours ensconced amid remote and windswept splendour, surrounded by tumultuous hosts of penguins, so I do not naturally gravitate towards captive animal exhibits. However, to give contemporary zoos and oceanaria their collective due, perspectives are changing. Generally, no longer mere menageries purely for public entertainment with little concern for animal wellbeing, their emphasis today increasingly focuses on developing sophisticated educational experiences. Though I can't say I've ever seen a person who doesn't end up amused by watching the lively antics of penguins, more important is the appreciation and conservation ethic the encounter often engenders. In this regard, I became impressed by how relaxed and happily busy all five species of true cold-climate penguins appeared inside San Diego's ultra-high-maintenance 'Penguin Encounter'. Indeed, the simulated habitat in this remarkable Antarctic re-creation offers engaging insights into several different species' real-world lifestyles, including a group of breeding Emperor penguins. With lighting that mimics south polar cycles, this hi-tech freezer facility, daily producing over 1.8 tonnes (4000 lb) of snow, permits visitors to gaze into an enthralling subzero penguin realm. Vignettes of interactions and behaviour — courtship, nesting, chick rearing and high-speed underwater chases — can all be observed in a real-time 3D sample of a world that very, very few people could ever be exposed to in true-life. The epigraph over the entrance says it all, 'highly specialized birds from a hostile but beautiful environment.'

Nowadays, about half of all the world's penguin species are in fact reasonably easy to see if one journeys

ABOVE AND BELOW Cute and endearing in their own wild environment, Magellanic penguins are notoriously pugnacious when defending themselves from intruders. Burrow nesting and forming dense colonies help them repel predators on the shores of Patagonia, Cabo Dos Bahias, Argentina.

to the appropriate southern hemisphere destinations. Even so, I find it indicative that more people have observed endangered wild tigers in India than have had the privilege of an encounter with Emperor penguins going about their business in the depths of Antarctica.

I believe the word you're looking for is 'Aaahh'!
— Oswald Cobblepot a.k.a. The Penguin, (Danny DeVito) in *Batman Returns* (movie), 1992

Not surprisingly, captive penguins represent one of the most popular zoo attractions worldwide. Undeniably, for the majority of non-travellers, such venues offer the only possibility of seeing live penguins. Thus, from London to Auckland, New York to Beijing, Dubai to Tokyo, Tenerife to the Philippines, hundreds of thousands of people from all walks of life flock to see a miscellany of penguin exhibits, some distinctly more elaborate than others. Generally featuring those species tolerant of warmer climates (Magellanic, Humboldt, African, Gentoo, Rockhopper, Macaroni, King), this multiplicity of so-called penguin-pools range, in my opinion, from educational and highly inspirational to woefully shameful, gimmicky attractions.

Significantly, with advances in seabird husbandry since the 1980s, colonies of all except the reclusive Yellow-eyed and several range-restricted crested penguins are successfully maintained in a number of specialised breeding facilities around the world. However, unlike some severely endangered mammal species whose future may hinge on captive 'stud' herds as viable back-up stock with which to repopulate the wilds, to date no penguin population has yet been pushed that close to the brink. Instead, the insidious threats to their long-term prospects are of major global proportions, involving shifting climates, dwindling habitats and increasingly stressed food resources — altogether bigger and more intractable issues than can possibly be addressed by sequestering a handful of breeders in zoos.

It's practically impossible to look at a penguin and feel angry.
— Joe Moore, American television personality

One facet predominantly in their favour is that by their very nature penguins make splendid ambassadors for themselves. They are simply immensely popular creatures; they uplift us; they make us smile. And it would appear that all around the world

people can readily identify with them. Or rather from whatever class, caste or rank, people seem to willingly identify penguins with themselves, comfortably anthropomorphising them into diminutive caricatures: the upright stance, the toddling gait, the crisp attire, their dignified personae and the élan with which we perceive them tackling life's challenges.

They're a charismatic embodiment of the famous aphorism: *Carpe diem...* but with panache.

Cults and culture

The nascent popularity of penguins began in the first decades of the 20th century. As coincidence would have it, exactly 100 years ago as I write this essay, a Norwegian whaling ship was bound for Scotland from South Georgia Island in the South Atlantic, carrying a cargo of valuable whale oil from the factories at Leith Harbour. Also stowed aboard were three freshly captured Kings, the first live penguins known to have been transported to the northern hemisphere. They were destined to become star attractions at the opening of the Edinburgh Zoo in January 1913, with a rather formal-looking likeness subsequently featured on the zoological society's Royal Charter coat of arms, as well as its more stylised modern logo.

Some 60 years later, this rather improbable Scottish–Norwegian–Penguin relationship took an unorthodox turn. In the early 1970s, an Edinburgh-raised King penguin was named Nils Olav, and then adopted as the official regimental mascot of the Royal Norwegian Guard. Still residing at the zoo in Scotland, Nils periodically inspected the foreign soldiers whenever they came to attend the famous Edinburgh Military Tattoo and, slowly, in an orderly succession of promotions, he rose through the Norwegian army ranks from Corporal in 1982, through Regimental Sergeant Major, to the distinguished echelon of Colonel-in-Chief. Then, in 2008, in a bizarrely unconventional military ceremony sanctioned by Norway's King Harald V, Nils Olav — now in his third incarnation after the initial draftee and his heir had each died of old age — was bestowed with a prestigious royal knighthood. Apparently on his best behaviour, and standing proud when the ceremonial sword was wielded around where his shoulders should be, the much-publicised accolade was performed before an enthusiastic crowd of spectators as Sir Nils appraised the parade of 130 of Norway's finest spit-and-polished guardsmen.

Military pageants notwithstanding, penguins — or, in most instances their effigies and similitudes — have

CREATIVECOMMONS.ORG ATTRIBUTION 3.0

ABOVE Mascot of the Royal Norwegian Guard, Sir Nils inspects the troops. A King penguin with officer ranking born and bred at the Edinburgh Zoo, in 2008 he became the only penguin ever to be officially knighted.

BELOW In the field of quantum physics a basic penguin diagram — a type of Feynman diagram here superimposed over a Chinstrap — represents the various interactive loop processes of elementary particle decay. CERN physicist John Ellis coined the term in 1977 after Melissa Franklin challenged him to include the word penguin in his research paper on bottom quarks.

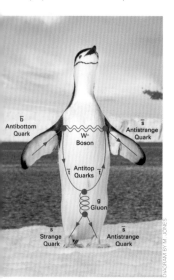

become as much a part of our diverse and modish lives as breakfast cereals (Nature's Path 'Penguin Puffs') and computer operating systems (Linux OS 'Tux' mascot – whose inventor, Linus Torvalds, was once bitten by a 'ferocious' Fairy penguin in Australia). From satirical, political-commentary comic book strips ('Opus' by Berkeley Breathed) to kids' online international playgrounds with over 15 million 7- to 14-year-old subscribers (Disney's 'Penguin Club'), notions and images of penguins have permeated our common psyche. On sports fields or in the halls of academia, infants to politicians, it appears the public has developed a veritable addiction to all things 'penguin'. Amongst a plethora of heterogeneities, the moniker features in a bewildering array of ways: on snack biscuits, ice-cream, designer clothing, video games and software, singing groups and orchestras, ice-hockey and lacrosse teams, a major publishing house, common household goods, a WWII Nazi bombing operation, a modern ballistic missile, numerous movies, cartoon characters, heroes, villains, awards, toys, books, stamps, coins, countless islands, Royal Navy ships (seven since 1757), innumerable private ships and yachts, a winery and even a town in Tasmania (population 4000). Just typing the word 'penguin' into the top online search engine may yield upwards of *175 million* hits. Meanwhile, your Boolean query has quite likely passed through a complex web-spam-thwarting protocol called the Penguin Update, data-refresh codes and enhanced algorithms that Internet giant Google unleashed in late 2012.

On the other hand, mention penguins to theoretical physicists working with gluons, quarks, bosons and fermions in the esoteric fields of particle quantum theory, and a very different image is instantaneously conjured. So-called Penguin Diagrams are used to illustrate such far-reaching concepts as 'electroweak penguin effects', 'penguin amplitudes' and 'penguin pollution', technical idioms conceived because the diagrammatical formulae and sketches of subatomic interactions are vaguely reminiscent of the shape of a plump, upright penguin. Admittedly — so the story goes — the penguin term was conceived in a whimsical eureka moment of desperation to win a bar-side bet between eminent CERN scientists in1977. It nonetheless graphically represents sophisticated

equations and notations that are beyond the grasp of all but the most analytical mathematical minds.

> *[the penguins] …were posed as though engaged in polite conversation at an afternoon garden party. Their self-absorbed congregation indifferent to the arrival of humanity….*
> — John Webber, artist on Cook's third voyage, of penguins on Kerguelen Island, December 1776

Whereas the arcane subterranean chambers of the Large Hadron Collider and the intellectual analysis of the deeper laws of nature may provide ample food for thought, to some of our ancestral peoples, penguins were simply food. Undoubtedly, the first humans to encounter penguins were the coastal tribesmen of southern Africa, the temperate seaboard Amerindians of South America and the southern Aboriginals of Australia and Tasmania. For millennia, penguins in those regions would have been harvested as little more than self-renewing prized commodities for their meat, their fat, their eggs and their useful skins. Latecomers onto this primal scene were the seafaring Polynesians, who only 1000 or so years ago voyaged southwards in the Pacific to colonise the uninhabited islands that would become New Zealand. There, they encountered an untouched bounty of rich and productive environments where they could hunt and fish and thrive. Alas, in the process, they brought about — or, for some species, at least initiated — the regrettable demise of a large number of unique native birds, among them a species of penguin, the enigmatic Waitaha (see Boessenkool, p.164). But then, like now, most penguin species favoured the inaccessibility of remote islands, where until relatively recent times, they remained beyond humanity's reach.

History and discovery

> *But look at the penguins of the Southern Ocean; have not these birds their front limbs in this precise intermediate state of 'neither true arms nor true wings?' …what special difficulty is there in believing that it might profit the modified descendants of the penguin, first to become enabled to flap along the surface of the sea … and ultimately to rise from its surface and glide through the air?*
> — Charles Darwin, *On the Origin of Species*, 6th ed., 1872

For once, Darwin was wrong: the penguin's underwater flight derived from air flight, not the other way around. All penguins evolved in the southern hemisphere, where their reconstructed phylogeny and biogeographical history implies that some 71 million years ago they shared a common volant ancestor with the albatrosses. From this, they diverged to follow their own evolutionary pathways towards not just simple flightlessness but rather to extreme and elegant adaptations to their marine existence. The ancient supercontinent of Gondwana had already begun to break apart, creating increasingly productive new oceans and seas; meanwhile, the mass

extinctions of the dinosaur era cleared the way for an explosion of new taxa; and the next 30 million years or so was generally a period of global cooling, all factors with major impacts on the evolution and proliferation of our modern biota. As the Antarctic continent shifted further over the pole and became ice-encrusted, and the other continents correspondingly drifted northwards, the developing circumpolar currents of the proto-Southern Ocean favoured the expansion of the prehistoric penguins. They were to become a diverse lot, and a good many long-extinct species – upwards of 50 – have been unearthed. The common ancestry of our modern penguins, beginning with the King and Emperor lineage, dates to about 13 million years ago, yet they differ remarkably little from the earliest known penguins (see Ksepka, p.158).

Although penguins are highly mobile birds, they have never crossed the thermal barriers of the equatorial regions to colonise northern latitudes. It therefore wasn't until European explorers and pioneers voyaged around the world in the 15th, 16th and 17th centuries that the narrative thread of people and penguins began to intertwine in a much more complex manner. Aboard the first caravel to venture into penguin country was a Portuguese navigator Diogo Cão, who reached the coast of Namibia in 1486. Yet, despite this being the northern range of African penguins, he made no mention of seeing them. Likewise, there was no reference by his compatriot, Bartolomeu Dias, who rounded Africa's 'Cape of Storms' in 1488 while pioneering the coveted direct sea route to the Spice Islands of the Indian Ocean. Consequently, it was in 1497 that an unknown sailor on Vasco da Gama's maiden voyage to India made the first direct written reference to birds 'as large as ganders and with a cry resembling the braying of asses', the sound which in due course would give rise to the African penguin's name of 'Jackass'.

> One day, when the air was filled with a sound of braying, sufficient to deafen one, I asked an old sailor… 'Are there asses about here?' 'Sir' he replied, 'those are not asses that you hear, but penguins.' The asses themselves, had any been there, would have been deceived by the braying…
> — Jules Verne, of Magellanic penguins in the Falkland Islands in *Le Sphinx de Glaces* (An Antarctic Mystery), 1897

To settle raging disputes and rivalry over prospective territorial claims, an absolutist Catholic decree resulted in the 1494 Treaty of Tordesillas, effectually dividing the whole of the unknown world beyond Europe into two equal shares between Portugal and Spain. Ergo it fell into the dominion of the Spaniards to pioneer the western sea routes towards the Pacific and beyond. This mission was commissioned to Fernão de Magalhães, an accomplished but out-of-favour Portuguese explorer auspiciously flying the Spanish Standard, better known today as Ferdinand Magellan. Though he didn't survive the voyage, his ship *Victoria* completed the first circumnavigation in 1522. More importantly to our story, before passing through

the straits that were to bear his name, Magellan spent several of the winter months of 1520 exploring the southeastern Patagonian coastline, where his prolific chronicler, Italian scholar Antonio Pigafetta, described the South American version of the Jackass penguin: … *we found two islands full of geese and goslings and sea wolves, of which geese the large number could not be reckoned; for we loaded all the five ships with them for an hour. These geese are black, and have their feathers all over the body of the same size and shape; and they do not fly and live upon fish; and they were so fat that [we] did not pluck them, but skinned them. They have beaks like that of a crow.*

And so we have our first detailed description of the Magellanic penguin, as well as the first of many accounts of the ransacking of penguin colonies; back then discovery and plunder went hand-in-hand.

Confounding etymology

> *The 24th of August we arrived at an island in the Straits, where we found great store of fowl which could not fly, of the bigness of geese; whereof we killed in less than one day 3,000, and victualled ourselves thoroughly therewith.*
> — Sir Francis Drake's Famous Voyage Round The World, Narrative by Francis Pretty 1577 (1910 edition)

In the eyes of early seafarers, geese, horses, asses, crows and fowl were common, oft-repeated analogies they used for many a strange new animal encountered on their voyages. For example, the first albatross were also described as goose-like, so it's easy to comprehend how confusions over the coining of certain names arose. Consequently it is from the sailors' sometimes oversimplified or colloquial lexicons of centuries past that we must elucidate the beginnings of the name we now know almost universally as 'penguin'. Shrouded in the folk-etymology of different languages and dialects, its roots have been alternatively associated with Breton, Welsh, Flemish, Norse, French, Spanish, Portuguese and Latin: 'Pen(n)' can mean 'head'; Gwyn (the source of a girl's name in several languages) or 'gouin' means 'white'; 'pinguis', from which we have the adjective 'pinguid' (meaning greasy), refers to fat or oil; 'pin-winged' has been debunked as fanciful English; and an obscure Germanic derivative, 'Fettgans', takes us full-circle to mean 'fat-goose'.

With so many ambiguous possibilities, the authentic etyma for the name have never been fully settled. But

ABOVE Watching penguins erupting from the sea to land on slippery rocks, such as these Gentoos on Cuverville Island, in the Antarctic Peninsula, it is easy to understand why Darwin thought this could represent the origin of flight in all modern birds. For once, he was wrong.
OPPOSITE BOTTOM The loud braying of Magellanic penguins confused seamen, thinking they were hearing asses on the shore, Saunders Island, Falklands.
BELOW A replica of Magellan's ship *Victoria* (the first ever to circumnavigate the globe) adorns the seafront at Puerto San Julián in the Patagonia region where, in 1520, his chronicler described the penguin that would eventually bear his name.

given that explorers tended to typecast new things after the old and familiar, it's not surprising to learn that their 'penguin' archetype was an entirely different bird, from the opposite hemisphere. Concurrently during the evolution of penguins, over the last 35 million years another very distinctive and unrelated group of birds emerged around the fringes of the northern oceans: the auks, auklets, murres, murrelets, guillemots, razorbill and puffins. With 22 living species in the family Alcidae, commonly known as auks, these became the highly specialised northern equivalents of the penguins in the south. Collectively they show moderate convergent evolutionary traits, with comparable lifestyles, in particular being accomplished pursuit divers using their wings for underwater propulsion. Except there is one great difference: all but one modern alcid retained the ability to fly. Consequently, perhaps, the largest of them today, the two murres, barely attain the size of the smallest penguin, the Little blue.

Naturally, humans have been interacting with alcids for much longer than with penguins — about as long as we have been a species capable of eking a living in the coastal environments of the higher latitudes. For millennia we've been harvesting them for food, oil and skins, still a common practice in many northern cultures today. There is evidence from archeological middens that even the Neanderthals already preyed upon them over 100 000 years ago.

The Great auk was the uniquely flightless giant of the group, standing at 75–85 cm (30–33 in) tall, upright like a penguin, and weighing over 5 kg (11 lb). Once ranging over the whole of the North Atlantic and a real hunter's prize, its similarity to some penguin species included relentless persecution as a basic commodity. As ships plied further and further into the remotest corners of the ocean, the magnificent Great auk was systematically extirpated. Finally, on 3 June 1844, the last known pair, from Eldey Rock off southwestern Iceland, were collected as museum relics.

A boat came in … laden with birds, chiefly penguins [Great auk]. … it has been customary [for] several crews of men to live all summer long on the island, for the sole purpose of killing birds… If a stop is not soon put to the practice the whole breed will be diminished to almost nothing, particularly for the penguins [Great auk].
— Captain George Cartright, 25 July 1785, from *Who Killed the Great Auk*, Jeremy Gaskill, 2000

Along with the Jackass (African) penguin, the Great auk had been one of the 564 birds initially described by Linnaeus, the father of zoological nomenclature, in his 10th *Systemae Naturae* of 1758. But the fallen giant left a series of remarkable legacies; mourning its loss, 19th-century conservationists were inspired to begin lobbying for the protection of various seabirds, including the Southern Ocean penguins, whose populations were by then similarly being over-exploited. The 'Last Great Auk' story joined the Dodo's to become classic textbook exemplars of how *not* to treat the natural world.

The most enduring — if confounding — part of the Great auk's story relates to its name. At some point in the early 1600s, its pursuers began referring to the tall, fat, flightless bird as 'pinguin', especially off the coast of Newfoundland, the species' final stronghold. Today, the French word *pingouin* remains the common name for all birds in the Alcidae family, whereas they call penguins in the south *manchots*. In 1791, Pierre Joseph Bonnaterre, a French naturalist expounding on Linnaeus' classifications, even nominated the Great auk to its own new genus, *Pinguinus impennis* ('fat penguin without wings'). In light of all this background, I find it not at all difficult to imagine an old, salt-worn sea-dog venturing for the first time into the southern hemisphere, declaring to his shipmates upon sighting an unknown, upright, black-and-white waddling bird plunge into the sea '…behold… a pinguin'… And so the Great auk, northern look-alike penguin, was revived into a new incarnation.

The sea is the element of the 'Manchot' [Spheniscidae]. Navigators often confuse them with the 'Pingouin' [Alcidae]; they differ in two distinctive characters, by the form of their wings, which, although very short and narrow in the 'Pingouin', at least allow them to rise and fly some distance; [and] by the configuration of the beak, which, in the 'Pingouin' is broad and flattened on the sides, whereas in the 'Manchots' it is pointed, rounded and cylindrical.
— Pierre Sonnerat, Naturalist aboard the *Ile de France* and the *Necessaire*, Voyage to New Guinea, 1776 (Translation from original French by M. Jones)

One of the most perplexing examples of confusing names is the Gentoo penguin, *Pygoscelis papua* — with singularly obscure origins to both its common and scientific names. Again, a quote translated from naturalist Pierre Sonnerat writing about the birds of New Guinea: '… I name three 'Manchot' I have seen: one the 'New Guinea Manchot' [apparently a King], two the 'Collared New Guinea Manchot' [after his drawing, likely a Magellanic], and three the 'Papuan Manchot' [a Gentoo is illustrated]'. It turns out that Sonnerat, while undoubtedly an enlightened and intelligent explorer, was perhaps a little liberal and imaginative in his prolific travelogues: apparently he didn't personally collect the penguin specimens, or in his many Pacific voyages did he

actually venture to the country we now know as Papua New Guinea! As for the puzzling common name — using the same seafarers' logic of naming something new after the familiar — Gentoo was perhaps coined in visual reference to the penguin's white headband, somewhat reminiscent of the white dupatta (traditional head scarf) of the cultures of India. Before the word Hindu had been widely accepted, 'gentue' (perhaps stemming from the original Portuguese word *gentio* for non-Christian 'gentile') was an archaic, later derogatory, term referring to all indigenous peoples of the Indian subcontinent.

Contributing further confusion to penguin nomenclature, 16th and 17th century scholars and naturalists making formal descriptions of new finds often had to rely on ship logs, journals, sketches and anecdotal commentaries for their information. Plus, some specimens were no more than decaying skins in jumbled collections of dubious provenance (in a later century, even Darwin mixed up some of his own collections). Nor did these forefathers of the life sciences have an intricate understanding of taxonomy and systematics, much less genetics, to call upon — for instance, until the 1750s even whales were regarded as fish not mammals. So it is that in 1781, Forster — the erudite naturalist from Cook's second voyage — officially described the Gentoo penguin from a type-specimen collected in the Falkland Islands, but, due to other erroneous accounts, blindly named it after somewhere no penguin of any kind had ever existed.

On the other hand, one penguin name that is indubitable was given in honour of Adèle Pepin, the much-adored wife of French explorer Jules Dumont d'Urville. In 1840, in claiming a newly discovered portion of the East Antarctica coastline since known as Terre Adélie (Adélie's Land), he wrote: 'This name is destined to recall in perpetuity, my profound acknowledgement of my devoted companion who has had to three times consent to long and painful separations in order for me to accomplish my faraway explorations.' In due course, the birds collected by d'Urville's naturalists were named after the land in which they were found.

Exploration and exploitation

The pace of new discoveries quickened and, in the recounting of more and more remarkable tales of lands, seas and islands, and especially of bounteous riches, those with a keen eye for profiteering were quick to follow in the wake of true explorers. So began the settling of colonial empires, the forging of trade alliances, and the rank exploitation and spoliation of natural environments. Lured by potential fortunes, these new waves of adventurers were mostly sealers and whalers (Palmer, Biscoe, Bristow, Weddell, Balleny and dozens of others) — courageous captains who methodically penetrated every corner of the southern seas, sometimes keeping their finds to themselves to forestall competition. Consequently, by 1854 all of the subantarctic islands, as well as the Antarctic continent — hence the complete gamut of penguin habitats — had been located, although five species, including the equatorial Galapagos penguin, had yet to be scientifically

described. In the fervent quest for fine oils and pelts to satisfy the demands of increasingly industrialised and opulent Europe and North America, whales were systematically hunted down and harpooned, fur seals and sea lions relentlessly clubbed, and elephant seals shot and butchered. Extravagant enterprises were developed around the promise of lucre and fame. In some cases within 10 years of a new island being investigated, the quarry was virtually exterminated. Not surprisingly, penguins didn't fare much better than the marine mammals: ships were loaded up with eggs, meat and skins, while crews living ashore for months at a time continually ransacked nesting colonies for food, even using spent penguin carcasses to fuel the fires of the constantly bubbling try-pots.

*The men said they had as much biscuit as they wanted, and also beans and pork, and a little molasses and flour. Their principal food was penguins (*Eudyptes chrysolophus)*, and they used penguin skins with the fat on for fuel...*
— H.N. Moseley, of Sealers on Heard Island in *Notes by a Naturalist: Observations Made During the Voyage of* H.M.S. Challenger, 1892

The most notorious case of devastation inflicted upon subantarctic wildlife — penguins in particular — was the story of one Joseph Hatch: erstwhile chemist, unforgiving merchant, artful politician, resourceful entrepreneur and unashamed capitalist. His enterprises, and the ensuing public outcries he engendered, also attest to the 20th century awakening a reversal in peoples' attitudes towards the treatment of wildlife. From 1862 onwards, Hatch — with the pithy business

ABOVE Hunted to extinction during the first half of the 19th century, the flightless Great auk, depicted in British ornithologist John Gould's landmark book *Birds of Europe* (left), was universally known to seamen as the 'Penguin'. Bearing a striking resemblance — even in its markings — to the Gentoo penguin (right), it is easy to see how the name was transferred.

BELOW The teeming throngs of King and Royal penguins lining the shores of Australia's Macquarie Island caught the attention of an infamous trader named Joseph Hatch, who soon transformed them into lucrative barrels of oil.

ABOVE Barrels of penguin oil are lined up next to a miserable hut on Macquarie Island, ready for shipment to lucrative markets in Europe and North America.

maxim 'More Production' — was, amongst his other schemes and interests, a zealous and remorseless sealer. Operating from his warehouses in Invercargill, at the southern tip of New Zealand's South Island (he also became the town's respected Mayor, and outspoken Member of Parliament), his sealing gangs methodically worked the thronging beaches of Auckland and Campbell Islands. Eventually, the New Zealand government, fearing total extermination of the animals, declared seasonal hunting bans from 1885 onwards. Hatch soon found himself severely disgraced for poaching during the closed season — a worthwhile risk, given that a single fur seal pelt was worth the tidy sum of 21 shillings on the London market (probably well over £150 using equivalent retail price indices today).

Having observed the megalopolis-like animal concentrations on Australia's Macquarie Island, with its 'multitudes of penguins and elephant seals', Hatch promptly refocused his efforts. With a ruthless eye for efficiency and high yields, he installed Norwegian-designed 'digestors', huge pressure boilers able to render entire 'sea elephant' carcasses, not just the blubber. When the lumbering pinnipeds became rare on Macquarie, he simply turned his attention to smaller animals, specifically the island's King and Royal penguins. While New Zealand authorities tried in vain to curb his brazen plunder (Macquarie was outside their jurisdiction) by heavily taxing his oil shipments, Hatch stepped up his enterprise with more and bigger digestors. By 1895 a single gang could render 2000 Royals *per day*, a figure that by 1905 had risen to 3500, with another 500 used to feed the constantly roaring fireboxes. While his men were often abandoned with scant supplies for months on end, he boasted that in a single season Macquarie generated 44% returns over his outlays.

...the most wretched place of involuntary and slavish exilium that can possibly be conceived; nothing could warrant any civilised creature living on such a spot.
— Captain Douglass, of the ship *Mariner*, commenting on Macquarie Island, 1822

Scoffing at scientific condemnation, Hatch retorted with written statements such as: 'I am prepared to prove that since I have worked the Macquaries, [penguins] have considerably increased ... [in] truth, unless a number... were slaughtered yearly, the birds would desert the island and disappear...'. After shifting his base to Tasmania in 1912, profits of the 'Southern Isles Exploitation Company' were temporarily bolstered by the unimpeachable needs for fats and fine oils during the 1914–18 Great War.

Meanwhile, the impatient 'Oil Baron' weathered increasing political and ethical criticisms but, in the end, his industry was doomed, forced into decline by public censure and ministerial pressure. With the backing of influential men of science, as well as the respected voices of accomplished and popular Antarctic explorers like Douglas Mawson and Frank Hurley, the wholesale mistreatment of Macquarie's penguins had triggered one of the first major international campaigns in defence of wildlife. With mounting pressure from London, Hatch's sole lease on the island was terminated, so his company became worthless and his equipment obsolete. By 1926 the brusque 89-year-old was bankrupt. He died two years later and, perhaps symbolically, lies in an unmarked pauper's grave in Hobart, while his digestors can still be seen rusting away amidst the steadily rebounding penguin colonies he annihilated in holocaust proportions: estimates suggest that some three million King penguins went through his oil refineries.

Changing attitudes

They are extraordinarily like children, these little people of the Antarctic world, either like children, or like old men, full of their own importance and late for dinner, in their black tail-coats and white shirt-fronts — and rather portly withal. We used to sing to them, as they to us, and you might often see a group of explorers on the poop, singing 'She has rings on her fingers and bells on her toes. . .' and so on at the top of their voices to an admiring group of Adélie penguins.
— Apsley Cherry-Garrard, quoting extract from Wilson's Journal, in *The Worst Journey in the World*, 1922

As the Hatch saga implies, the first decades of the 1900s marked a significant turning point in the relationship between people and penguins. Not only were those first captured King penguins beginning to entertain crowds in Edinburgh, but newspaper and lecture hall audiences were becoming enthralled by the odysseys of a patriotic new breed of adventurer. The so-called Heroic Age of Antarctic exploration was in full swing, with over a dozen expeditions penetrating ever deeper into the frozen south.

Edward Wilson was a zoologist on both of Robert Falcon Scott's famed Antarctic expeditions of 1901–04 and 1910–13. So inspired was he by Darwin's misguided hypothesis that Emperor penguins were the most

BELOW Monuments to a gloomy past, the hideous digestors of the Hatch enterprise on Macquarie Island stand mute amid the recovering colony of King penguins at Lusitania Bay. One hundred years ago, five such pressure plants were dotted around the island.

126a. Sky effect (midnight sun), penguins at ice-edge. Jan. 13th 1911.

primitive of birds, whose embryos should shed light on evolutionary links with reptiles, that he and two companions set out on a mid-winter quest for specimens from which they were lucky to return. The five-week ice-trek in total darkness from their base hut at Cape Evans on Ross Island to the only known Emperor colony at Cape Crozier became an epic tale of survival, of battling ferocious storms and -40°C temperatures, all to collect Emperor eggs at a critical stage of incubation. Wilson and Bowers tragically perished, along with Scott, during the return journey from the South Pole the following summer. It fell to his other companion Apsley Cherry-Garrard to deliver the samples to London's British Museum of Natural History. There, the three precious eggs that three men had toiled to secure languished unstudied for 20 years, during which time the embryo hypothesis was refuted.

Another intriguing twist to Scott's last expedition came from team scientist George Levick's account of the 1911–12 Adélie penguin breeding cycle at Cape Adare, where he was the first, and to this day the only, scientist to study the colony throughout an entire season. Levick's detailed behavioural observations were far ahead of their time. So much so that his accounts of the penguins' frequent sexual activities and particularly of obsessive 'hooligan bands' of males (truculent young birds that would forcibly attempt to copulate at random) were so shocking to Edwardian sensibilities that he encoded passages in Greek so only scholars should read them. Deemed too distasteful and indecent for public eyes, extracts were declined for publication in the expedition's official reports. These chronicles of 'astonishing depravity' were set down in a pamphlet entitled *Sexual Habits of the Adélie Penguin*, then lost amongst original expedition records for nigh on 100 years. Rediscovered during preparations for the Scott Centennial Celebrations, upon review by modern scientists his amazingly accurate descriptions match those of the most recent studies.

Feeling somewhat like a giant who had wandered on to the wrong planet, and who was distinctly in the way of its true inhabitants.
— Apsley Cherry-Garrard, in *The Worst Journey in the World*, 1922

But of all of Scott's men it was Herbert Ponting, expedition photographer and cinematographer, who made the most exoteric and lasting contribution to the future of penguins: he unabashedly popularised them. Starting out as a freelance photojournalist, he immortalised the atmosphere and trials of polar life with singular sensitivity. His glass plates and motion pictures have left us a haunting legacy of that fateful expedition. Equally outstanding were his many public presentations over subsequent years, illustrated by magic lantern (a projector-like device) and delighting audiences with stories and images of the penguins he'd lived alongside. Some of his penguins appeared as prints or as postcards, and five photographs of Adélies he presented to England's King George V in 1914 still reside in the Royal Collections. But Ponting's

pièce de résistance was, in what was arguably the first application of motion picture merchandising, his invention of a small stuffed Adélie penguin replica, crafted from fur fabric, velvet and cotton and named Ponko (his own nickname on the ice). This first ever 'cute and cuddly' toy penguin became the popular mascot of his tours, engaging children and adults alike. Recently recast as the central character in a children's educational storybook series about Antarctica, the original Ponko now lives in retirement at the National Maritime Museum at Greenwich.

Penguin appeal

Although we may have come a long way in our attitudes towards penguins during the last 100 years, one perspective has not changed all that much since Hatch's day: penguins, in all their many guises, are definitively a good business resource. Big business.

The paragon of the marketability of the penguin label began in 1935 when it was adopted as a 'dignified but flippant' symbol by Penguin Books founder Allan Lane, in a modest attempt to make the then extortionately priced 'intelligent books' available as cheap paperback editions for the lay public. In launching what was to become one of the most universally recognised brand names of all time, he commented that '… these Penguins are the means of converting book-borrowers into book buyers…'. Initially treated with suspicion by his traditionalist contemporaries, Lane's ethos transformed the publishing world: within two years over three million readers had a Penguin in their pockets and for a sixpence were able to buy the latest coveted paperback title directly from a London high street vending machine called the Penguincubator. In 2010, the company's 75th birthday, standing as the world's second largest publishing house, annual revenues were above £1 billion.

A disparate contemporary of the Penguin paperback was the superhero comic book, the stage for one of the most enduring anthropoid penguins ever invented. In 1941, artists Bob Kane and Bill Finger gave birth — figuratively speaking — to Oswald Chesterfield Cobblepot, a bullied child of short, rotund stature, gross weight and beak-like facial features. Not everyone may know these names, but the fictitious genius Cobblepot is well recognised as the perennially inventive, umbrella-wielding Batman adversary, 'The Penguin', who remains one of DC Comics' most popular villainous cranks.

While the jerky-motion Chaplin-era silent movies of Ponting's Adélies may have been viewed with inquisitive enthusiasm, they were barely the trailer for the penguins that were to hit the screens in later years.

ABOVE LEFT Herbert Ponting, official photographer and cinematographer on both of Scott's Antarctic expeditions, promoted penguins like never before.
ABOVE Using postcard advertisements featuring 'Ponko', his signature Adélie penguin, Ponting attracted huge crowds to his lectures, films and photo shows.
BELOW Paragon of polar exploration literature, Apsley Cherry-Garrard's landmark adventure book of Scott's 1910–13 expedition was popularised in 1937, becoming the 100th publication under the revolutionary Penguin Books trademark.

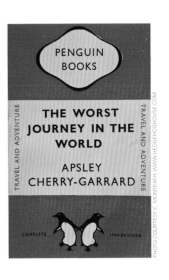

PHOTO COURTESY OF MARITIME NEW ZEALAND

ABOVE Spewing heavy bunker fuel, in 2011 the grounded container ship *Rena* spelled disaster for penguins and other seabirds of the Bay of Plenty, New Zealand.

PHOTO COURTESY OF ROBYN SHARROCK

ABOVE An innovative project in Australia uses specially trained dogs to protect Little blue penguins from introduced predators.

BELOW Because of high risks to the environment, ships that depend on heavy bunker oil for fuel, like this passenger-carrying icebreaker, are no longer permitted in Antarctic waters.

From the early 1930s right through to the present day, we have evolved some very sophisticated and popular penguin stars, for many of which Walt Disney's talented cartoonists paved the way. On 1 September 1934, a nine-minute Silly Symphony cartoon movie short entitled *Peculiar Penguins* debuted in full Technicolor, depicting in classic Disney song-and-dance fashion the happy-ever-after love story between two cutely animated penguins. Ever since, playing alongside icons like Donald Duck, Mickey Mouse, Goofy, Bugs Bunny and Woody Woodpecker, all manner of fictional penguin-styled caricatures have besieged the marketplace. Of course, personal favourites tend to be the ones we grow up with, so Chilly Willy, who reigned through the 1950s and 1960s, is perhaps my pick of the bunch. But among the dozens of penguin animations that have come in and out of vogue over the years, Pingu, a pre-school children's claymation (stop-motion modelling) character has transcended international language and cultural barriers like no other. Developed in 1984 for Swiss television, and described as 'very different... quirky, warm and lots of fun', Pingu's BAFTA award-winning antics and real-sounding (regardless of your mother-tongue), totally improvised 'penguinese' dialogue, has been watched by over a billion people on over 140 television stations worldwide, generating multi-million dollar profits on the plasticine penguin's consumer products.

Meanwhile on the big screen, James Bond may be invincible, but a bunch of larking computer-generated penguins gave the suave, tuxedoed spy a smart dressing-down at the box office. In November 2006, *Happy Feet* tap-danced around *Casino Royale*, pipping the gallant agent to top-grossing spot during the films' joint opening weekend, with the penguins raking in a cool $41m in debut screenings. As a prime example of penguin-movie appeal, by the end of 2012, *March of the Penguins* (2005) remained the all-time top grossing nature documentary.

Into the new millennium

Watching a wildlife documentary can often be the catalyst for people to want to experience exotic nature first-hand. Penguin-centred tourism has become a competitive travel sector in recent years, and encompasses anything from camping in the Falkland Islands or guided wilderness hikes to remote beaches in Westland, New Zealand, to boat rides and picnic trips to Magellanic and African penguin colonies, or grandstand-like night-time viewing platforms to watch homecoming Little blues (see Where to see penguins, p. 236).

The holy grail of penguin destinations is, of course, a fully-fledged expedition-ship cruise to Antarctica and the diverse subantarctic islands. Although the chequered history of southern exploration, coupled with the majestic scenery, are obviously important themes for Antarctic travellers, it's the attraction of seeing penguins that determines most itineraries. In fact, in the Antarctic Peninsula segment, 67% of the 142 specified visitor sites south of 60 degrees latitude (the international legal boundary of the southern continent) are at penguin breeding colonies. Extrapolating from the published 2011–2012 season data from the International Antarctic Tour Operators Association, of the 20 most visited locations some 86% of tourist landings are at places that feature nesting penguins, with only 24 sites (including all the research bases and historical artefacts) that don't feature them at all. In that season, over 26 000 visitors, of more than 100 nationalities, had their own Antarctic penguin fix.

With all this discussion around what penguins mean to people, and of our emotional ties with these captivating little beings, what of the penguins themselves? Travel as we may, we can actually witness very little of their real lives as many species spend up to 75% of their time at sea. Only recently, with the advent of satellite tracking devices, have we been enlightened on just how far, and to where, many species routinely travel (see Wilson, p. 170; Wienecke, p. 176).

Today it is impossible to think of animals without concern for man's impact upon them.
— Bernard Stonehouse.

Recent conservation-minded generations have actively tried to right the wrongs we have inflicted on penguin colonies in years past. By legislating for reserves and government-protected areas, eradicating introduced pests that adversely affect breeding success, and proactively enhancing and restoring the habitats of species that live in close proximity to human activity, many previously decimated species are in better shape nowadays than even just a few years ago. Some solutions, while labour intensive and costly, are comparatively easy to implement with the appropriate resources, whereas others require more innovative thinking. One award-winning initiative at a Little blue colony on Middle Island, near Warrnambool, about

260 km (160 miles) west of Melbourne, Australia, has, in a very short time, been particularly successful. Foxes and wild dogs had ravaged the colony of some 600 penguins until only about 10 remained. In 2006 Maremma sheepdogs, a 2000-year-old Italian breed with strong protective instincts — first used to ward bears and wolves away from domestic sheep and goat herds, but in Australia effectively employed to protect chicken farms — were specially retrained as guardian dogs for the penguins. As of the 2012 summer census, approximately 190 breeding birds were again resident on the island. These days, with the dogs so efficiently keeping the foxes at bay, one of the biggest challenges of the project is to prevent the dogs becoming bored!

But in spite of our best actions at some locations, a number of species are still at risk, whether from marauding predators or altogether more complex and less manageable threats. Of particular concern is petroleum pollution, both from general shipping and from oil drilling and its affiliated operations. Many maritime high-traffic areas overlap with penguin habitats, with inevitable dire consequences. In 2011 alone, two fully laden cargo vessels ran aground near penguin nesting areas — one on Nightingale Island in the mid-Atlantic (see Glass, p. 188), and another sailing between ports in New Zealand — spilling large quantities of heavy bunker fuel oil that killed large numbers of penguins and other seabirds. Both were totally avoidable navigational errors where officers' negligence drove their ships onto well-charted obstacles. These accidents merely serve to highlight that new offshore oil prospecting operations in prime penguin territories such as the Falkland Islands, Argentina, Chile, Peru, South Africa, South Australia and around New Zealand, will in time bring additional major catastrophes. For instance, a single oil spill anywhere off New Zealand's Banks Peninsula — an area recently gazetted for drilling — could potentially wipe out the entire white-flippered subspecies of Little blue penguins. Arguments for the unlikeliness of such events occurring can be countered with figures pulled from the International Tanker Owners Pollution Federation statistics: in 42 years of worldwide records, 1350 spills of over 7000 L (1500 gallons) have been logged, plus a further 455 major spills over 700 tonnes. Granted, since more stringent regulations have been implemented, tanker accident averages since 2002 are less than half those of the preceding decade, and still declining, but the fact remains that even a single spill, especially one associated with Big Oil, can be all too devastating. In South Africa, a specialist non-profit organisation called The Southern African Foundation for the Conservation of Coastal Birds (SANCCOB) dedicates itself full-time to the rehabilitation of distressed seabirds, the majority due to oil. Having treated over 90 000 African penguins, their efforts have been rewarded by seeing a wild population some 19% higher than it would otherwise have been. These represent the kind of efforts that are necessary to stave-off dramatic human-induced declines. SANCCOB is now a world leader in oiled-bird rescue, a depressingly frequent happening along the current-swept coasts of southern Africa and elsewhere. Just as it was for the oiled Northern rockhoppers of Tristan da Cunha, no doubt their expertise will be called upon again and again in future (see Ryan, p. 182; Boersma, p. 184).

A gradually spreading awareness and growing concern for the overall health of our world's ecosystems is slowly permeating the global human intellect. Our planet is a dynamic entity of which we are all intrinsically and intricately a part. While some of our individual activities and their effects may appear to be easily quantified, in reality they are anything but. We can collectively accept or deny human responsibility for the root causes; either way, it does not alter the fact that our climate *is* changing, our oceans *are* becoming more acidic and their primary productivity *is* declining; that glaciers and icecaps *are* shrinking, and annual sea ice coverage *is* dramatically reduced. Already the differences can be measured, and to mitigate them will require a paradigm shift in human behaviour (see Ainley, p. 166; Lynch, p. 168).

A tenet of communication theory is that a person is more likely to make influential decisions about matters that reinforce personal values. What can be of greater value than trying to safeguard the planet's ecosystems? We all have a vested interest, and we all have a responsibility to act appropriately. Penguins, those charismatic little creatures-turned-social phenomena, whose positive influence on people is so great that we seem to be unable to live without them, are at the heart of these issues. Their very existence — and ours — depends on the future welfare and resilience of the oceans. In the following pages, some of the world's leading penguin experts recount their personal stories of research and discoveries, trials and tribulations, and share some intriguing facets of penguin biology and enigmas that have been brought to light, in the process expressing their deepest concerns over the future of many penguin species.

ABOVE Changing attitudes: an old Antarctic base encroaches directly into a Gentoo penguin colony in Paradise Bay (right) while passengers from an expedition ship watch King penguins from a respectful distance on South Georgia (left).

ABOVE Where King penguins and their eggs were once harvested, tourism now grants them priority, Volunteer Beach, Falkland Islands.

BELOW It will take more than a traffic sign to give penguins right-of-way in New Zealand coastal cities.

March of the Fossil Penguins

Daniel Ksepka

Dr. Daniel Ksepka is a Curator at the Bruce Museum. His research focuses on the early history of penguin evolution and includes work on penguin phylogeny, neuroanatomy, histology and biogeography.

Bruce Museum Greenwich, CT 06830 USA ksepka@gmail.com

Synopsis: Penguins represent an ancient avian lineage that reverted to flightlessness early in the Cenozoic Era. Rebuilding their family tree from the fossil record brings their story into focus.

Penguins are an ancient group of birds. Primitive species first appear in the fossil record over 60 million years ago. Early chapters in penguin evolution are the primary focus of my research. Pursuing fossil evidence of their ecology and evolution has brought me to many localities where fossil seabirds occur in abundance, such as New Zealand, Peru and South Africa.

Over the course of their evolution, penguins have experimented with shifts in body size, morphology, diet and physiology. More than 60 extinct species are known from fossils, many representing fascinating side branches of the penguin family tree that were successful for a few million years but ultimately died out without leaving any descendants.

Primitive penguins inhabited a world quite different from today. During the Eocene Epoch (36–56 million years ago) the southern supercontinent of Gondwana was still breaking apart, and Australia and South America were in close contact with Antarctica. Climate was much warmer as well, a phase recognized as a 'Greenhouse Earth' interval with no permanent ice sheets at the North and South Poles. Marine mammals like dolphins and seals were absent during the first stages of penguin evolution, which may have allowed particularly large penguins to explore some niches that are not open to them today.

A central aim of my research has been to reconstruct an evolutionary tree for penguins past and present, because this tree can serve as a crucial 'roadmap'

for evolutionary biology. DNA evidence can ascertain the relationships of the living species of penguins, yet understanding these relationships alone is like stripping a family tree of all the grandparents and great-grandparents — so much of the story is lost.

As I proceeded to place the fossils into their proper position on the family tree, some surprising implications emerged. One is that the oldest fossils of modern penguins (those that shared the most recent common ancestor of all living species) are only 11 to 13 million years old, suggesting that turnover between archaic species and modern species happened quite recently from a geological perspective. Another is that the distribution of fossil penguin lineages strongly suggests that early species originated in temperate latitudes. Despite being associated with icy environments in the popular imagination, penguins appear to have colonized permanently glaciated environments very late in their history.

Fossils also reveal penguins started evolving some key thermoregulatory structures like the countercurrent heat exchangers in their flippers — which efficiently transfers arterial heat to returning venous blood flow with minimal heat loss to the extremities — long before this period. Such features can be thought of as exaptations, structures that evolved for one purpose and were later co-opted for another. In this case, thermoregulatory structures that first evolved for conserving body heat on long underwater foraging trips 50 million years ago are now one key for surviving the icy Antarctic breeding season in the modern world.

Where do penguin fossils come from? It turns out that many parts of the world that penguins live in today have fossil localities nearby, testifying to the deep history of these birds. To walk along rocky outcrops where marine deposits are now above sea level is to literally walk along the ancient ocean floor. It is amazing to look up from this prehistoric marine environment and find yourself surrounded by miles of Peruvian desert or the gently rolling hills of a New Zealand pasture. Sandstones,

ABOVE TOP Remarkably intact bones of the fossil Peruvian penguin *Incayacu paracasensis* enable a reconstructed diagram of a 1.5-m long giant that lived around 36 million years ago. ABOVE BOTTOM The 36-million-year-old skull of a spear-beaked giant, *Icadyptes salasi* from Peru, compared to an extant Magellanic penguin. RIGHT Paleontologist Dr Ewan Fordyce of Otago University examines a composite skeleton of the slender *Kairuku grebneffi*, a fossil species he discovered in southern New Zealand, hotbed for penguin evolution.

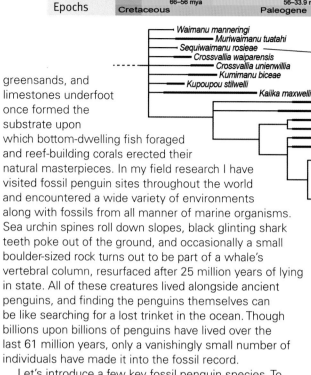

Paleogene / Neogene

ARTWORK COURTESY SIMONE GIOVANARDI

Waimanu manneringi
Muriwaimanu tuatahi
Sequiwaimanu rosieae
Crossvallia waiparensis
Crossvallia unienwillia
Kumimanu biceae
Kupoupou stilwelli
Kaiika maxwelli
Notodyptes wimani
Perudyptes devriesi
Anthropornis grandis
Anthropornis nordenskjoeldi
Palaeeudyptes gunnari
Palaeeudyptes klekowskii
Burnside "Palaeeudyptes"
Inkayacu paracasensis
Icadyptes salasi
Pachydyptes ponderosus
Kairuku sp.
Kairuku grebneffi
Kairuku waitaki
Glen Murray Kairuku
Archaeospheniscus lopdelli
Archaeospheniscus lowei
Paraptenodytes antarcticus
Platydyptes marplesi
Platydyptes novaezealandiae
Eretiscus tonnii
Palaeospheniscus bergi
Palaeospheniscus biloculata
Palaeospheniscus patagonicus
Marplesornis novaezealandiae
Nucleornis insolitus
Aptenodytes ridgeni
Aptenodytes forsteri
Aptenodytes patagonicus
Pygoscelis grandis
Pygoscelis tyreii
Pygoscelis adeliae
Pygoscelis antarcticus
Pygoscelis papua
Madrynornis mirandus
Eudyptula minor
Eudyptula novaehollandiae
Inguza predemersus
Spheniscus muizoni
Spheniscus megaramphus
Spheniscus urbinai
Spheniscus chilensis
Spheniscus demersus
Spheniscus magellanicus
Spheniscus humboldti
Spheniscus mendiculus
Megadyptes antipodes waitaha
Megadyptes antipodes antipodes
Megadyptes antipodes richdalei
Eudyptes atatu
Eudyptes moseleyi
Eudyptes chrysocome
Eudyptes filholi
Eudyptes chrysolophus
Eudyptes schlegeli
Eudyptes pachyrhynchus
Eudyptes robustus
Eudyptes sclateri
Eudyptes warhami

Crown clade

greensands, and limestones underfoot once formed the substrate upon which bottom-dwelling fish foraged and reef-building corals erected their natural masterpieces. In my field research I have visited fossil penguin sites throughout the world and encountered a wide variety of environments along with fossils from all manner of marine organisms. Sea urchin spines roll down slopes, black glinting shark teeth poke out of the ground, and occasionally a small boulder-sized rock turns out to be part of a whale's vertebral column, resurfaced after 25 million years of lying in state. All of these creatures lived alongside ancient penguins, and finding the penguins themselves can be like searching for a lost trinket in the ocean. Though billions upon billions of penguins have lived over the last 61 million years, only a vanishingly small number of individuals have made it into the fossil record.

Let's introduce a few key fossil penguin species. To date, the oldest known taxa are the ~60 million-year-old 'proto-penguins' *Waimanu manneringi*, *Muriwaimanu tuatahi*, and *Sequiwaimanu rosieae* of New Zealand. Although they were clearly flightless, these species had less flattened wing bones and could still fold their wings to a certain extent. These early penguins appeared soon after one of the most important events in Earth history, the mass extinction at the end of the Cretaceous Period. This extinction, believed to have been caused by an asteroid impact, wiped out the non-avian dinosaurs, along with many other groups of animals. It is possible this extinction benefited early penguins by removing predators and competitors such as the marine mosasaurs and plesiosaurs.

New Zealand was home to many different fossil penguin species, including the gigantic 35 million-year-old *Pachydyptes ponderosus*, long known as one of the largest penguin species ever to have lived. Recently, an even larger species was discovered. The 57 million-year-old *Kumimanu biceae* tilted the scales at ~120 kg (265 lb). Some unusually svelte penguins also plied the ancient seas of New Zealand including my own personal favourite species *Kairuku waitaki*. This penguin had elegant proportions, with a straight elongated beak, slender trunk and long, narrow flippers. At over 1.2 m (4 ft) tall, these birds would have cut a striking figure on New Zealand's beaches 27 million years ago. The Chatham Islands, located 800 km (500 miles) east of mainland New Zealand, have the sad distinction of being home to one recently discovered penguin species that was wiped out by humans — the crested penguin *Eudyptes warhami*, which was driven to extinction 700 years ago.

Peru is another hot spot for penguin evolution. Although only the Humboldt Penguin lives in this region today, a rich array of fossil species are recorded in ancient ocean sediments that are now exposed in the deserts of the Pisco and Ica regions. During the Eocene Epoch, penguins were living close to the Equator in Peru — at a time when the global temperature was on average 5° to 8°C hotter than today. Some species, such as *Icadyptes*

salasi, were giant spear-billed penguins adapted to taking large prey. This drives home the variety of environmental and morphological adaptations seen in early penguins. Another remarkable Peruvian discovery was a 'penguin mummy' belonging to the species *Incayaku paracasensis*. At a glance, the feathers of this penguin appear fairly modern, with the small flat-shafted feathers forming overlapping scale-like layers on the wing. But on closer inspection, preservation of pigment-bearing organelles called melanosomes indicate that *I. paracasensis* likely possessed a coat of grey and reddish-brown feathers, unlike any adult penguin today (see pages 162–3).

Paleontologists are only starting to scratch the surface of early penguin evolution. We have entered a modern 'golden age' of discovery, with more than a dozen new species described in the last decade. With more attention focused on fossil sites and ancient penguin research than ever before, more fascinating insights into ancient penguins are surely in store for us in the next few years. ⚲

ABOVE Timeline of penguin evolution.

LEFT One of the heaviest known penguins was the colossal Pachydyptes ponderosus. This species lived in New Zealand ~35 million years ago. Its massive humerus (at right, compared to the same wing bone from Emperor, Adélie and Little Blue penguins) suggests it weighed 65kg (143lb).

PHOTO COURTESY J.C. STAHL, MUSEUM OF NEW ZEALAND TE PAPA TONGAREWA

How Penguins Store Food: The Discovery of an Antimicrobial Molecule

Yvon Le Maho

With an academic background in physiology and main interest in understanding how animals operate under natural conditions, Dr Yvon Le Maho has been studying penguins for more than 40 years. His work transcends major disciplinary boundaries, bringing together researchers from widely differing fields to elucidate some extraordinary penguin adaptations that yield new potentials in medical science.

Institut Pluridisciplinaire Hubert Curien Département Ecologie, Physiologie et Ethologie UMR 7178 CNRS-Université de Strasbourg, 23, rue Becquerel F-67087 Strasbourg Cedex 2, France yvon.lemaho@iphc.cnrs.fr http://iphc.cnrs.fr

Synopsis: A remarkable discovery reveals evolutionary adaptations in incubating fasting male King penguins: capable of retaining food in their stomachs for weeks to feed to newly hatched chicks, they fend off bacterial invasion with a unique antimicrobial peptide, which carries exciting potential for biomedical sciences.

For many pelagic seabirds one of the inherent challenges of breeding is finding ways to cope with the alternating regimes of feeding at sea and fasting ashore during egg incubation and chick-guarding phases. In some species, like Emperor and King penguins, the distances

ABOVE After seven weeks of incubation, the single chick emerges from the egg balanced on top of its father's feet.
RIGHT The male, who usually takes the final 2–3 week incubation shift, may feed the newly hatched chick for several days if his mate has not yet returned, relying on well-preserved food stored in his stomach, a trait not found in females.

required to find their prey translate into foraging trips lasting several weeks. Accordingly, when mates swap duties, it is obviously critical that the one remaining in the colony possesses enough body fuels to last the duration, as well as have sufficient food remaining in its stomach to feed the newly hatched chick.

In the 1990s, my PhD student Charles Bost, using Argos satellite transmitters, found that the King penguins breeding at Possession Island in the Crozet Archipelago rely primarily on myctophid (lantern) fish that concentrate along the Polar Front, the shifting oceanographic boundary where subantarctic waters meet the colder Antarctic seas. To reach this productive mixing zone, the penguins must cover distances varying between about 300 km (185 miles) in a cold year and up to 600 km (370 miles) during warmer periods, climatic variations essentially linked to the El Niño Southern Oscillation phenomenon.

Stimulated by the research of Yves Handrich, we raised the interesting question: what happens when such warming occurs while the male King penguin assumes the final two to three weeks of incubation and the well-provisioned female's return — which normally coincides with the time of hatching — is delayed by a week or more due to the increased travel distance? We found that in the female's absence, the male is still capable of feeding the chick to ensure its survival, without himself returning to sea to fish. And, that he can do this for as long as 10 days or so. This means the male has conserved food unspoilt in his stomach since coming ashore to replace the female for the last incubation shift. Our paper on this finding, in which Michel Gauthier-Clerc played a key role, caused quite a sensation when it appeared in the Christmas 2000 issue of *Nature* magazine.

In light of these observations, the next question we wanted to resolve concerned the mechanisms for preserving such food undigested while the birds are fasting ashore for up to three weeks. We all know that fish cannot be kept for long, even at temperatures much lower than a penguin's body, and that the toxins in natural bacterial degradation would likely cause the death of either parent or chick. So, in 2001, I decided I should seek answers to how the penguins successfully avoid this fate.

At that time, I was following the work of Jules Hoffmann, who was leading a research institute in

The results were striking. Compared to the stomach bacterial flora of non-breeders during normal digestion, in males saving food towards the end of incubation we found the bacteria were either dead or appeared to be stressed, supporting the idea of some antimicrobial mechanism in operation. King penguins have a gastric temperature of 38°C and a pH of about 4, clearly ideal acidic conditions that would favour bacterial proliferation, so such a mechanism would indeed be necessary for conserving food almost intact over several weeks.

Over more than two years, by separating different fractions in the stomach content samples, we isolated an antimicrobial peptide (small protein), which appeared under two parent molecules. Based on the scientific name of the family of penguins, Spheniscidae, we called it spheniscin (Sphé-1 and Sphé-2). It belongs to the family of defensins that are involved in the intrinsic immune defence of mammals and birds. From what we already know regarding the mechanisms implicated in the activity of defensins against bacteria, they make the development of bacterial resistance more difficult than with antibiotics. Thus, several properties of spheniscin suggest that it might have a particularly high potential in biotechnological and biomedical applications.

All the structural and biological data we obtained on the two spheniscin molecules supported their role in food conservation and in the protection of the penguin stomach epithelium (lining) against pathogenic bacteria and fungi. In particular, we found that they are especially efficient against the bacteria *Staphylococcus aureus* and the fungus *Aspergillus fumigatus*, two main agents of nosocomial diseases, otherwise known as hospital super-bugs. Moreover, the stomach content being similarly saline to our ocular environment (in other words, to our tears), spheniscin is, as far as we know, unique as a defensin in remaining active in such a saline matrix. Added to the fact that to date physicians have an insufficient panoply of antimicrobial substances to fight ocular infections in humans, the possibilities for a medical breakthrough of significant proportions — all coming from penguins living in the far southern reaches of our planet — are indeed momentous.

How would I have ever thought, when starting my research more than 40 years ago, that some of our data would result in such potential applications for human health. While the full development of new biomedical substances takes years, spheniscin promises to join the vast ranks of invaluable biodiversity services on offer, and which we certainly would never have discovered had King penguins disappeared about a century ago when they were being boiled down for their oil by the tens or hundreds of thousands throughout the Subantarctic. ⏏

Strasbourg investigating the antimicrobial molecules produced by insects — the so-called defensins — as I thought similar molecules might be involved in food conservation in penguins' stomachs. Jules, who about 10 years later was to receive the Nobel Prize in Physiology and Medicine for this research, offered me the opportunity to work with Philippe Bulet, his expert on antimicrobial molecules. Funding for this work remained a big issue because I was stepping outside my field of expertise as a penguin biologist, seeking to venture into new frontiers of immunology, molecular biology and proteomics. Fortunately, Hubert Curien, father of the Ariane Space program and former Minister of Research, who had recently been appointed chair of a committee in the Foundation of France, rescued the project. This enabled me, in addition to my field work in the Crozet Islands, to hire the postdoctoral candidate Cécile Thouzeau to follow through with laboratory research for 2–3 years.

King penguins incubate their egg on their feet for about 55 days and it seems that during this period only the males engage in saving and conserving food in their stomachs, and this only during the last shift of incubation. Obviously, with the close association of food conservation while fasting tied to a particularly sensitive phase of incubation, it was critical to avoid any disturbance which could lead to abandoned eggs. Fortunately, we had the essential field expertise to take several small samples of stomach contents from the sitting kings without jeopardising their breeding success.

LEFT Timing her return from the sea around hatching, a female prepares to take over the days-old chick from the male's care.
ABOVE AND BELOW Emperors may store food in their stomachs for weeks while travelling back to feed their chicks, but fasting males at the end of the two-month winter incubation actually produce a protein-rich stomach secretion akin to milk for their hatchlings if the females are late returning.

Penguin Colours and Pigments: A Few Surprises

Matthew Shawkey

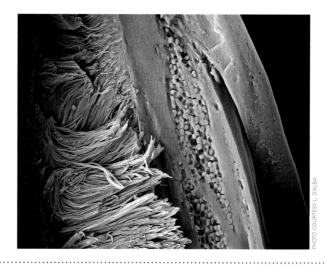

PHOTO COURTESY L. D'ALBA

Dr Matt Shawkey was first introduced to birds by his high school biology teacher and for the past 12 years has studied the optics and evolution of bird colours as a PhD student at Auburn University, postdoctorally at the University of California–Berkeley and now as associate professor of biology at the University of Akron.

Associate Professor Integrated Bioscience Program University of Akron Akron, OH 44325-3908 USA shawkey@uakron.edu http://gozips.uakron. edu/~shawkey/

ABOVE Unique light-scattering nanofibres, not pigment, are responsible for the blue colour in Little blue penguin feathers.
BELOW AND RIGHT Ultraviolet light is reflected off the bright beak patches of Kings (left) and Emperors (right), possibly used to convey their fitness for mate selection.

Synopsis: Behind the stereotypic image of penguins as plain black-and-white birds, several studies reveal unique colour-producing mechanisms and pigments in feather and beak adornments, plus an unexpected find in a 36-million-year-old fossil.

We typically think of penguins as having relatively simple black-and-white colouring. This classic tuxedo styling is more than simply striking; rather, it likely serves as a camouflage. A penguin blends in with the dark of the ocean floor when viewed from above and with the light of the sun when seen from below. This so-called 'countershading' probably helps to disguise them from both predators and prey, a strategy employed by many other seabirds and marine mammals.

However, it turns out that penguins did not always have this colour pattern. My collaborator Jakob Vinther and I have recently discovered a way to scientifically reconstruct the colours of fossil organisms using data from microscopic pigment-containing sacs called melanosomes that are sometimes present in well-preserved specimens. We determine how size and shape of melanosomes is related to colour in modern birds, and then apply these relationships to fossilised melanosomes, enabling us to predict colour.

In 2010 my colleague Julia Clarke and her research team discovered a new species of extinct giant penguin on the Paracas Peninsula, along the desert coast of Peru, later named *Inkayacu paracasensis* or roughly 'water emperor from Paracas'. Amazingly, this 36-million-year-old ancestor of modern penguins not only showed well-preserved bones, but actually held fine details of feather structures, including microscopic melanosomes.

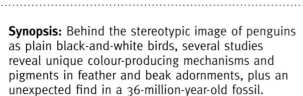

We decided that this presented a perfect opportunity to learn about the evolution of colour in penguins. Our comparisons between melanosomes from this ancient giant penguin and those of modern birds yielded two surprise findings. First, melanosomes from *Inkayacu* were smaller than those from modern penguins, and more similar in size and shape to those of other birds. Apparently, modern penguins have evolved larger melanosomes, perhaps to increase stiffness of feathers. Second, *Inkayacu*'s plumage lacked countershading. Instead it had a brown

underside and grey back. That the primary predators of penguins, such as seals, were diversifying at this time suggests that the shift towards countershading may have been an evolutionary response to increasing predation pressure.

Several modern penguins, however, still show interesting variations in their plumage colour beyond simple black and white. Little blue, or Fairy, penguins have a unique blue tint to the dark portion of their plumage on their heads and backs. Postdoctoral Research Associate Liliana D'Alba and I became intrigued by this colour while examining penguin pigmentation during the Inkayacu project, and decided to investigate what creates this deep bluish tinge. Using various microscopy and optical techniques, we found that the colour is produced not by blue pigments but by orderly arrays of microscopic fibres about 200 times smaller in diameter than a human hair. The size and arrangement of these nanofibres causes scattering of light that leads to blue light being reflected from the feathers. While similar arrays of parallel collagen fibres were known from avian and mammalian skin, this mechanism of colour production is entirely new in feathers. Other birds create bright blue colours in their plumage through air-filled, sponge-like arrangements of keratin that coherently scatter light, causing selective reflection of blue wavelengths. Whether the blue colour serves an adaptive function in this penguin or is involved in mate selection are questions that remain to be answered.

Further exciting discoveries in penguin coloration have been made in different research. Kevin McGraw and his co-workers revealed that the yellow head feathers of several penguin species, such as the plumes of Macaroni and other crested penguins, contain a unique fluorescent pigment not found in other bird species. They first noted that these feathers fluoresce in a yellow-green colour under a UV lamp, then verified these properties by separating the pigments by centrifuge and quantifying fluorescence emission with a scanning spectro-fluorometer. In fact, not only did this

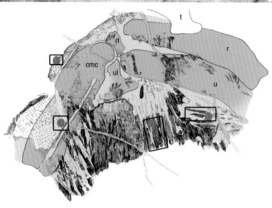

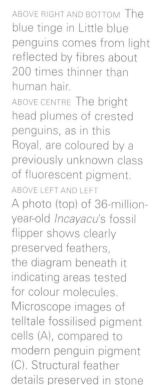

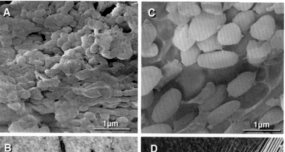

ABOVE RIGHT AND BOTTOM The blue tinge in Little blue penguins comes from light reflected by fibres about 200 times thinner than human hair.

ABOVE CENTRE The bright head plumes of crested penguins, as in this Royal, are coloured by a previously unknown class of fluorescent pigment.

ABOVE LEFT AND LEFT A photo (top) of 36-million-year-old *Incayacu*'s fossil flipper shows clearly preserved feathers, the diagram beneath it indicating areas tested for colour molecules. Microscope images of telltale fossilised pigment cells (A), compared to modern penguin pigment (C). Structural feather details preserved in stone (B) appear identical to a modern emperor penguin feather (D).

study uncover a previously unknown pigment, but it appears to represent an entirely new class of pterin pigments, unlike the more common carotenoid- and melanin-based plumage colours in most birds. Whether this fluorescence is visible to penguins in natural conditions is unknown, but it certainly makes their sometimes outlandish headgear quite striking to our eyes, and may serve as a form of advertisement to potential mates.

In a subsequent study, the same author was able to neatly correlate the intensity of yellow colour visible in Snares-crested penguins — caused by high concentrations of this pigment — with body condition in the largest, heaviest, healthiest-looking individuals, especially females.

Another interesting study by Pierre Jouventin et al showed that brightly coloured patches on the beaks of King and Emperor penguins had high reflectance properties in the ultraviolet wavelengths. Although all adults possess this trait, the team found that the highest reflectance levels were in recently paired birds. Unlike humans, most birds can see in the UV spectrum, suggesting that these markings may be a way for penguins to communicate with one another, perhaps revealing their age, breeding status or fitness qualities, without becoming more conspicuous to non-avian predators.

These examples demonstrate that, even under the constraints of their marine existence, unusual colour schemes can play significant roles in the lives of birds not typically

Waitaha Penguin: Dynamic History Revealed by DNA

Sanne Boessenkool

Dr Sanne Boessenkool is a researcher at the Centre for Ecological and Evolutionary Synthesis, University of Oslo, Norway. Her research focuses on retrieving DNA from ancient and (pre) historic specimens to elucidate changes in species and population distributions.

sanneboessenkool@gmail. com

ABOVE AND BELOW The Yellow-eyed penguin took the place of the mysterious Waitaha along the shores of New Zealand's South Island after humans arrived.

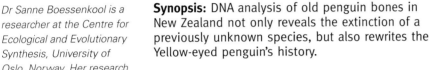

Synopsis: DNA analysis of old penguin bones in New Zealand not only reveals the extinction of a previously unknown species, but also rewrites the Yellow-eyed penguin's history.

I vividly remember the moment that a colleague and I were looking at the very first results of our DNA analyses of old penguin bones from New Zealand. At first, we thought it looked messy and there was no pattern at all, but within minutes we started to realise that we were staring at something rather unexpected!

In the last decades, technological advances have made it possible to study the DNA from specimens that are hundreds, or even thousands, of years old. By comparing this DNA to that from individuals alive today, we can learn, for example, how abundant populations have been at a certain point in time, whether they have increased or declined, and sometimes even if specific individuals have migrated. For my research, I had set out to study the effect of human settlement in New Zealand on the Yellow-eyed penguin (*Megadyptes antipodes*), an endemic and endangered species with approximately 1800 breeding pairs alive today, and an icon of the country's conservation management.

Since many bird species have severely declined or even become extinct since Polynesian and European settlement in New Zealand, I had expected Yellow-eyed penguins to have also been much more abundant in the past. In order to test whether this was indeed the case, we compared DNA from 1000-year-old penguin bones originally excavated from the archaeological ovens of the first Polynesian settlers with the DNA of 100-year-old museum specimens, as well as contemporary samples. But what we discovered was completely contrary to our expectations.

The older bones from the New Zealand mainland were not actually Yellow-eyed penguins at all. Instead, they belonged to a new species that until then had remained completely unnoticed to science. Analyses of these bones by palaeontologists Trevor Worthy and Paul Scofield revealed that not only was their DNA different, but also their shape and size confirmed their uniqueness. With that, the discovery of a new species became a fact,

and we named it *Megadyptes waitaha*, or the Waitaha penguin, in reference to the earliest pre-European inhabitants of New Zealand.

Unfortunately, the Waitaha penguin no longer survives. It likely went extinct, along with many large terrestrial flightless birds, within a few centuries of the first humans arriving in New Zealand from the Pacific Islands around 1000 years ago. Such extinctions are not surprising, considering that these early colonists

depended primarily on hunting for their livelihood. Archaeological ovens found in the remains of their villages typically contain vast amounts of animal bones, both terrestrial and marine, including those of the Waitaha penguin. Penguin meat has a high fat content so would have been of high nutritional value. Large island species are particularly vulnerable to intense hunting as a consequence of their relatively slow reproductive rates. Additionally, penguins are generally curious and approachable animals, making them quite an easy catch, especially as they had never been exposed to mammalian land predators prior to humans.

We know very little about the Waitaha penguins. There seem to be no drawings of them or any oral history of their existence, as there are for some other extinct New Zealand species. The size of the Waitaha bones shows that these penguins were a little smaller than the Yellow-eyed penguin, and the DNA tells us they were very closely related. But whether they also had the significant yellow eye or the yellow stripe across the head remains a mystery.

Strangely, the extinction of the Waitaha penguin had a positive spin-off for the Yellow-eyed penguin, which became much more abundant and spread to the New Zealand mainland. Our findings reveal that before humans had arrived in New Zealand, Yellow-eyed penguins lived only on the subantarctic Auckland and Campbell Islands. Undoubtedly every now and then an individual successfully crossed the approximately 500-km (300-mile) stretch of ocean separating these islands from the New Zealand mainland, but while the Waitaha penguin existed they had not managed to establish themselves permanently. This occurred only after the Waitaha's extinction. The geographical range expansion of Yellow-eyed penguins is quite remarkable, and the obvious question is, of course, 'what had changed?', such that this spread was suddenly possible. There may have been quite strong competition between the two species, preventing the Yellow-eyed penguin from settling where the Waitaha was already established. Furthermore, one of the Yellow-eyed's main predators, the New Zealand sea lion, had also been hunted by the pioneer people until it was no longer present on the New Zealand mainland. Without this predator patrolling the beaches, the place perhaps became a little safer for Yellow-eyed penguins. Apart from these environmental changes, we also think there may have been a gradual shift in the developing culture among the Polynesian settlers, after which the penguins were no longer hunted.

It seems that the expansion of Yellow-eyed penguins took place well before New Zealand was colonised by Europeans, who brought with them a variety of mammalian land predators. These introduced species (e.g., mustelids such as stoats, ferrets and weasels, and cats) soon turned wild and currently exert great predatory pressures on penguins. Today only 442 pairs have been censused around the South Island of New Zealand, and another 200 or so for Stewart Island.

All in all, it appears that a unique combination of favourable conditions allowed the Yellow-eyed penguin population to grow and expand its range from the subantarctic islands to the New Zealand mainland at a particular juncture in time. From a conservation perspective, perhaps the most important message is that in the event the mainland population becomes locally extinct, there is no guarantee that this colonisation process could ever be repeated.

The idea that the Yellow-eyed penguin could still become extinct on the New Zealand mainland is not far-fetched. Currently, its population abundance is unstable and numbers have fluctuated strongly over recent decades. It is thought that besides predation by introduced predators, changes in food supply, climatic variations and disease epidemics all contribute to these fluctuations; a combination of factors very difficult to control through conservation management. To try and elucidate these matters further we have carried out additional research on current levels of genetic variation and the extent of migration among present-day Yellow-eyed penguin populations. Low genetic variability has revealed that the mainland population was established by only a small group of founders from the Subantarctic, with very limited interchange between the two gene pools. Consequently, the mainland population may have only a limited ability to adapt to environmental changes (e.g., the emergence of new diseases), which could explain the instability we observe today.

In conclusion, the DNA of prehistoric and contemporary penguins in New Zealand has told us not only of the dynamic history of the Yellow-eyed penguin, but has also revealed an even richer prehistoric fauna, sadly adding yet another species to the list of human-induced extinctions. 🐧

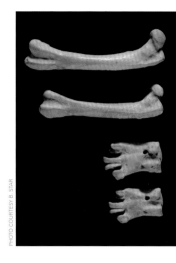

PHOTO COURTESY B. STAR

ABOVE Yellow-eyed and Waitaha penguin leg bones clearly show the smaller size of the latter.
BELOW A Yellow-eyed penguin among flowering megaherbs on subantarctic Enderby Island, the species' original habitat.
OPPOSITE Private penguin reserve, with habitat restoration and protection from predators, set up on a private sheep farm, Penguin Place, Otago Peninsula.

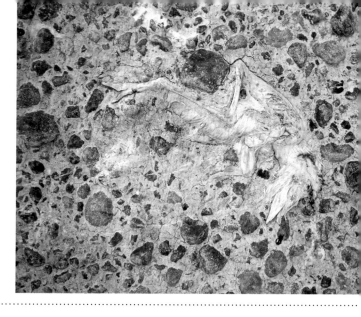

Adélie Penguin: Bellwether of Ecosystem Change

David Ainley

Dr David Ainley is a marine ecologist, having gained a PhD in ecology from Johns Hopkins University in 1971. He has been investigating Antarctic top predators since 1968, including seven stints in multi-investigator oceanographic cruises, nine times across the Drake Passage, four trips between the Ross Sea and New Zealand, three cruises into the Weddell Sea, plus 29 summers at Cape Crozier/ Cape Royds, Ross Island, and two at Anvers Island, Antarctic Peninsula investigating Adélie and emperor penguins, plus Weddell seals, and their prey.

Senior Associate Ecologist H.T. Harvey & Associates Ecological Consultants 983 University Ave, Bldg D, Los Gatos, CA 95616, USA dainley@penguinscience.com www.penguinscience.com

ABOVE AND BELOW Mummified carcasses trampled into the frozen ground, and piles of nesting pebbles, both used to identify the presence of past colonies.

Synopsis: Adélie penguin's remarkable adaptability to environmental change is evident by combining real-time observations with ancient records in marine and land deposits.

The Adélie penguin is no stranger to altering ecosystems, especially with respect to climate change. Throughout its existence, likely extending from early Pleistocene (2.5 million years ago), it coped with four major ice ages, each with attendant advance and retreat of glaciers and sea ice. Owing to Antarctica's cold, dry conditions preserving penguin bone and eggshell remains for thousands of years, a unique record exists, untold for any other species. In places, even their little piles of nesting stones are still recognizable thousands of years later. As a result, the Adélie penguin has been called 'the bellwether of climate change.'

To understand this relationship, consider that the species is an obligate associate of sea ice, occurring only where pack ice is a major feature for at least a few months of the year. Adélies can deal with an ice-covered ocean, especially by an ability to rapidly accumulate body fat to live off while fasting until the ice loosens to allow water access. Also, they can hold their breath longer for their size than other penguins (up to 6 minutes, but usually 2 to 3), enabling them to reach prey beneath the ice itself. On the other hand, too much sea ice, especially in the form of extensive, unbroken, land-locked sheets called fast ice, compromise their ocean access and ability to reproduce. If they must walk long distances, rather than swim, between their nesting and feeding grounds, the energetics may lead to complete nesting failure.

Only two regions of Antarctica's coastline align roughly north–south rather than east–west: the Antarctic Peninsula and Victoria Land, the western shore of the Ross Sea. Studies of these areas illustrate how Adélies deal with the long-term, climate-induced advance/retreat cycles of sea ice. During the Little Ice Age, a period 450–800 years ago when Earth's overall climate cooled significantly, Adélie colonies spread northward along the western Peninsula. We know from ocean sediment cores that this matched an increased, northward persistence of pack ice in the area. Nowadays, however, the western Antarctic Peninsula, due to vagaries of the polar jet stream, is warming rapidly. Concurrently, Adélie penguin colonies and sea ice are fast withdrawing southward.

Research along Victoria Land indicates that thousands of years ago similar movements occurred at the southern extent of the penguin's range, but for a different reason. During an unusually warm period 4000–5000 years ago, Adélie colonies existed along the coast of southern Victoria Land. But with subsequent slight cooling, and the return of extensive fast ice, those colonies disappeared and have not been seen since. Ancient nest bowls, eggshell fragments and bones are all that remain.

Going back in time to the last Ice Age 12,000– ~27,000 years ago, with the glacial maximum about 18,000 years ago, the only continental shelf not ice-covered — during this or in any previous Ice Age — was the northeast corner of Victoria Land. Seafloor sediment cores reveal that winds kept the adjacent ocean ice free, allowing Adélie penguins to breed on a volcanic headland called Cape Adare. As a result, the Ross Sea population became isolated; today they are genetically distinct from all other Adélies, whose Ice Age stock derives from those inhabiting more northerly islands that escaped severe glaciations. At the start of the Holocene, Ross Sea Adélies followed the retreating ice front, and by 9000 years ago began to reoccupy locations where they are found today, deep into the Ross Sea. Moreover, strata of colony remains even tell us of their cyclical presence and absence going back some 45,000 years.

These findings were derived primarily from careful reconstruction of existing evidence, but without knowing the demographic process involved. Then, in 2001, a remarkable event provided a 'natural experiment' that

in essence duplicated the return of an Ice Age on a local scale, when a huge sliver of the West Antarctic Ice Sheet broke off and lodged against Ross Island. At 165 km (102 miles) in length (the size of Jamaica), this iceberg, named B15 by glaciologists, presented an impassable barrier to penguins returning to breed deep in the Ross Sea that spring. Besides forcing them to detour, it also trapped a huge area of sea ice, preventing its normal breakout by the time growing chicks require fast food deliveries. Of our four study colonies, the three eastern ones (Capes Royds and Bird on western Ross Island, plus offshore Beaufort Island) were cut off, while the fourth and largest at Cape Crozier remained accessible.

For the six seasons prior to this event, we had been banding 2000–5000 chicks annually among the four sites, keeping detailed records on their location and breeding status when at an older age they returned to the colonies. We had noted a tendency for a few to emigrate from Crozier to the smaller colonies, where there was less competition for space and food, but with their southbound route essentially blocked this suddenly changed. Rather than trying to circumvent the obstacle, many of our known Royds and Bird penguins turned up at Crozier and stayed. Those that did make it to Royds that season had to commute on foot 70 km (43 miles) over ice. All those attempting to breed eventually deserted their nests, while the subadults didn't even visit. During 2–3 years of failure, about half of Royds adults moved north to Beaufort Island and Cape Bird, where the distance to open water was shorter. Interestingly, when in 2005 a small tsunami floated the iceberg off the subsea pinnacle on which it had perched, few of these emigrants returned to their original colonies even though the sea ice pattern of the region returned to normal.

During this time, another change was happening that demonstrates the adaptations that come with relaxation of the Ice Age. Owing to slightly warming air temperatures, the glaciers on Beaufort Island began to retreat, opening up nesting space, thus easing the pressure on chicks raised here to emigrate to other colonies. We knew this from banded fledglings at the Ross and Beaufort islands. When more habitat became available on Beaufort, emigration to other colonies ceased.

However, another twist emerged out of this grand-scale climate experiment. As the Crozier colony grew from approximately 170,000 to over 300,000 pairs between 1996 and 2019, chick growth decreased, compared to the other smaller colonies: average fledging mass dropped almost 1 kg (2 lb). We know the distance

parents can effectively travel to secure food for their chicks is normally limited to about 100 km (62 miles), but as the Crozier colony grew, competition for food forced parents to travel as far as 150 km (93 miles). Their trip frequency dropped and chicks were fledging at suboptimal weights. It appears that the 'climate refugees' swelling the Cape Crozier ranks effectively were eating themselves 'out of house and home.' We deployed an ocean glider with acoustic sensors by which the 3-D presence of food in the penguins' foraging area could be quantified. We found, indeed, that as the chick-feeding period progressed the penguins were depleting prey close to the colony.

Ironically, despite fledging at lower and lower mass, thus potentially compromising subsequent survival, the Crozier colony, as was also true for Cape Bird and Beaufort Island, continued to grow. It seemed, therefore, that adults and fledglings were finding plenty of food beyond the colony's range constraints. Seeing that they were also eating more Antarctic silverfish (higher nutrient value than krill), we hypothesized that the commercial fishery for Antarctic toothfish (sold as Chilean sea bass), that began in the Ross Sea in 1997, was depleting one of the penguins' main competitors. More silverfish available would allow the penguins to compensate for their prey depletion near to colonies.

In conclusion, while under stable conditions Adélie penguin 'law' is to return almost unfailingly to their natal colony to breed, they readily change strategy when conditions change. As today's Ice Shelves break up and disappear along the Antarctic Peninsula, we can expect northern Adélie colonies to continue to vanish while they occupy new locations exposed by ice loss to the south. If current climate change trends continue, and surpass 2°C above pre-industrial (1850) levels, the models we have generated predict that only the Ross Sea — the southernmost stretch of ocean on the planet — will retain the sea ice that this ice-loving penguin requires. But other ice-dependent species such as Antarctic toothfish, silverfish and krill will also change their range, with unknown consequences for Adélie penguins. 🐧

ABOVE Southernmost of all penguin colonies is on Cape Royds at 77°S.
BELOW LEFT Adélie colonies in the Ross Sea appeared and disappeared in tandem with pulsating ice cover over the last 40 000 years, as the Ross Ice Shelf grew and shrank in response to climate oscillations.
BELOW RIGHT Track of the enormous B-15 iceberg that affected the fate of nesting colonies in the Ross Sea during several years.
BELOW OPPOSITE With roughly 250,000 nesting pairs, Cape Adare supports the largest stable Adélie colony.

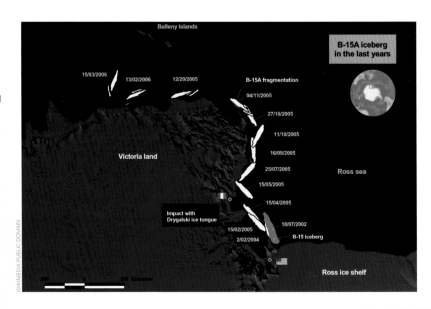

Spying on Climate Change: Monitoring Antarctica's Penguins from Space

Heather J. Lynch

Dr. Heather J. Lynch is the IACS Endowed Chair for Ecology & Evolution at Stony Brook University. Her research focuses on the distribution and abundance of Antarctic wildlife, combining field work with satellite imagery. She and colleagues have developed the Mapping Application for Penguin Populations and Projected Dynamics (www. penguinmap.com). Dr. Lynch has a degree in Physics from Princeton University and a PhD in Organismic and Evolutionary Biology from Harvard University.

Professor, Ecology & Evolution Stony Brook University Stony Brook, NY 11777, USA heather.lynch@stonybrook.edu www.Lynchlab.com

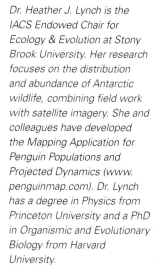

ABOVE RIGHT Hypothermic Adélie chicks huddle together when unusually high air temperatures turn snowfall to rain, South Orkney Islands.
RIGHT By developing algorithms to detect penguin colonies from satellite imagery, huge new colonies of Adélies were discovered in the ice-choked Danger Islands near the tip of the Antarctic Peninsula. Numbers were later confirmed by creating detailed mosaics from drone photography, such as Beagle Island with 284 535 nesting pairs.

Synopsis: The integration of traditional field surveys and high-resolution satellite imagery is driving a paradigm shift in biogeographical studies of Pygoscelis, or brush-tailed, penguins in response to climate change around the Antarctic Peninsula.

Like the proverbial man looking for his keys under the street lamp because that's the only place he could see, penguin biologists have historically been confined to studying penguin populations within easy reach of a research station. These long-term studies have provided a wealth of information on reproduction and survivorship, but remain few and far between, scattered as they are over a continent the size of the United States and Mexico combined. An alternative approach, one we've used for decades, is to survey populations opportunistically from research or commercial cruise vessels; while passengers visit penguin colonies, we count all the nests or chicks we can to assess the size of the breeding population in that year. This approach provides economical access to hundreds of locations and has been enormously informative to our understanding of the Antarctic Peninsula's brush-tailed penguins (Adélie, Chinstrap, and Gentoo, *Pygoscelis* spp.).

However, counting penguins this way still leaves major gaps in our understanding of the total population. Passenger vessels only visit the most accessible sites along the Antarctic Peninsula, leaving the rest of the continent untouched. The largest penguin colonies can take days or even weeks to count, but with cruise ships we usually only have a few hours in which to complete our census work. Such challenges are universal among Antarctic penguin biologists and, as a result, the largest penguin colonies often go unstudied. But no more.

Although penguin biologists have been trying to use satellites (and before that, airplanes) to study penguin colonies since the technology first became widely available [Schwaller et al 1984], the recent proliferation of high-resolution satellite imagery has created something of a gold rush in the field of penguin biology. Advances over the last decade have been dramatic and personally exciting for the growing cadre of specialists working at the interface of

satellite image interpretation and Antarctic ecology. My own introduction to this work came in 2010 at the International Penguin Conference, over a cup of coffee with Paul Morin and Michelle LaRue from the University of Minnesota's Polar Geospatial Center. I had spent several years during my PhD using Landsat imagery to map insect outbreaks in Yellowstone National Park and, having shifted attention to Antarctic penguins, was intrigued by the possibilities afforded by higher-resolution commercial satellite imagery. Paul and Michelle were involved in a collaboration with the British Antarctic Survey to use satellites to survey Emperor penguins that, by virtue of their large size and visual contrast (black-backed birds on white ice), were an obvious target for this kind of technology. As one cup of coffee turned into two and then three, it quickly became clear that by integrating data collected in the field with satellite information, we could start surveying not just the remote colonies on the Antarctic Peninsula, but eventually all the penguin colonies in Antarctica.

In the twelve years since that first meeting, the dream of mapping every penguin colony in Antarctica

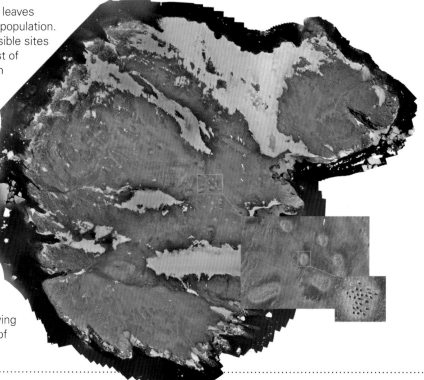

has come to reality. Both high-resolution commercial satellite imagery such as Worldview and Quickbird, as well as medium-resolution sensors such as NASA's Landsat series, have proven invaluable tools to fulfill this work. Satellites have been critical in facilitating the first global population surveys of Emperor penguins [Fretwell et al 2012], Adélie penguins [Lynch and LaRue 2012], and Chinstrap penguins [Strycker et al 2020a], as well as the first island-wide survey of King penguins on South Georgia [Foley et al 2020].

Satellite-based surveys yield estimates that are highly comparable to field counts and yet require a tiny fraction of the time and money required for ground-based efforts. Freed from the logistical constraints imposed by distance, ice and dangerous oceans, satellites have allowed us to estimate populations in some of the world's most remote places, like the South Sandwich Islands in the South Atlantic [Lynch et al 2016] and across the distant coastlines of Eastern Antarctica [Lynch and LaRue 2012]. Not only do satellites replace a lot of difficult and expensive field work but, by allowing us to see penguin colonies and the surrounding landscape ahead of a field expedition, they make work on the ground safer as well. Satellite imagery played a key role in planning major expeditions to the southwest reaches of the Antarctic Peninsula [Casanovas et al 2015] as well as to Elephant Island [Strycker et al 2020b], a craggy island once covered in Chinstraps, made famous by Shackleton's infamous journey back from the brink of disaster.

One of the most exciting satellite-based discoveries was our accidental finding of a series of very large penguin colonies in the Danger Islands, an ice-choked archipelago off the northwestern tip of the Antarctic Peninsula. NASA scientist Matt Schwaller and I had trained an algorithm to identify penguin guano in Landsat satellite imagery [Lynch and Schwaller 2014], which at 30 m (98 ft) resolution was much coarser than the sub-metre quality pictures we'd been using but was, paradoxically, easier to interpret. This algorithm had flagged a large area of guano in a region not known to host breeding penguins, so I went in search of higher-resolution imagery to resolve the algorithm's 'problems'.

It turns out that the computers had been correct all along! There were penguins all over the Danger Islands and some colonies appeared to be among the largest in Antarctica.

Armed with these maps, we teamed up with other penguin researchers to organize an expedition to the Danger Islands, where we found penguins as far as the eye could see! Unmanned aerial drones provided the perfect perspective needed to survey these massive colonies and, in the end, we discovered that this tiny archipelago contained more Adélie penguins than the rest of the Antarctic Peninsula region combined [Borowicz et al 2018]. As a result, this formerly neglected island chain is now widely recognized as a major penguin hotspot worthy of formal protection, which is a nice illustration of how better technology really can lead to better penguin conservation.

But satellite imagery has provided more than just population counts. We can use the colour of penguin guano to extract information about their diet [Youngflesh 2018], track the movement of Emperor penguin colonies navigating a shifting icescape [LaRue et al 2015], and watch Gentoo penguins establish new colonies in the shadows of melting glaciers [Herman et al 2020]. We can even use the shapes of penguin colonies to study the health of the population, because colonies that are stable or growing are rounder and less reticulated than those suffering declines [McDowall and Lynch 2019]. In this way, satellite imagery is contributing not only to our understanding of penguin numbers, but to our understanding of penguin behaviour.

The future is bright for using satellites to study penguins, and over the next few decades we will be able to watch penguin populations grow and shrink and move in response to environmental changes in a way we could only dream of in the past. Climate change on the Antarctic Peninsula has outpaced global averages by a wide margin, and such rapid warming has had major impacts on the sea ice [Stammerjohn et al 2008] and the food web [Clarke et al 2007]. While satellites can't stop climate change, we now have a window through which to watch how the penguins are coping. ♦

TOP The color of penguin guano, which varies whether fish or krill predominates in their diet, can be seen on satellite photos, revealing health trends in penguin populations.
ABOVE Active island arc volcanoes, the South Sandwich Islands harbour a major proportion of the world's Chinstrap penguins.
LEFT (BOTH) A census obtained from satellite views of the large Bailey Head colony are consistent with ground counts, but can be conducted in a fraction of the time.

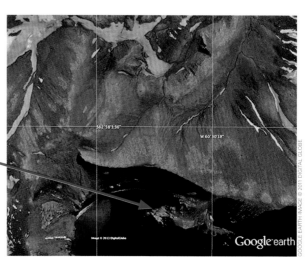

Beyond Prying Eyes: Tracking Penguins at Sea

Rory P. Wilson

Born in the United Kingdom, Dr Rory Wilson read zoology at Oxford University before undertaking a PhD on the foraging ecology of the African penguin at Cape Town University. Specialising in elucidating the marine behaviour of penguins, with over 300 published peer-reviewed scientific articles, he is currently a professor in Aquatic Biology at Swansea University.

Swansea Lab for Animal Movement Biosciences, College of Science Swansea University, Singleton Park, Swansea SA2 8PP, Wales, United Kingdom R.P.Wilson@swansea.ac.uk

ABOVE Until recently, our ability to study penguin behaviour stopped the moment the subjects entered the water. RIGHT The latest technology, some of it invented by the author, shown opposite with one of his subjects, reveals detailed information on individual foraging trips, e.g. this 3D Magellanic penguin dive profile off the coast of Argentina. FAR RIGHT Learning where African penguins have been when they return to feed their chicks provides data needed to better manage resource competition from fisheries.

Synopsis: A life-long curiosity about penguin behaviour at sea has prompted some inventive technology, revealing many surprises about their diving performances.

When I started my PhD on the foraging ecology of the African penguin in 1980, I had no idea what I was letting myself in for. A severe population decline was believed to be due to problems at sea, so my job was to gather data on penguin feeding habits. I remember very clearly my first day on Marcus Island walking along the exposed coast, watching these unlikely heroes as they plunged into the roaring surf and promptly disappeared. Squinting into the sun on the water, I recall thinking, I'm in big trouble.

I discovered a number of important things in the first few weeks on the island. Firstly, that it is almost impossible to determine much about penguin behaviour by watching them at sea from boats. Foraging penguins spend most of their time underwater, so observation from above is not terribly helpful. Secondly, any attempt at gaining meaningful information by watching them underwater is immensely frustrating. It is possible to see them predictably as they swish past close to their colonies, but aside from demonstrating their stunning speed and agility, this tells us next to nothing about the ways of the real penguin using the depths and vagaries of the ocean to make a living.

The solution then, and now, was to attach some instrumentation to the birds to record what they did

at sea. I played with various harness designs and a number of different homemade gadgets. Though crude compared to today's digital technology, with these basic devices I soon found that African penguins could dive as deep as 130 m (425 ft) — though they mostly do not exceed 30 m (100 ft) — and could swim at almost 20 km/h (12 mph) in pursuit of small oceanic schooling fish. The information was simple but revolutionary, and threw a whole new perspective on the capacities of these exceptional birds.

Knowing the maximum depth wasn't enough for me. I spent years tinkering with systems that could be attached to penguins' backs to record what they did during their mysterious sea sojourns. I progressed to using tiny radioactive beads to expose sensitive film stuck next to handmade penguin speed-meters to discover that African penguins cruise underwater at about 9 km/h (5.5 mph) and generally forage for chicks within 20 km (12.5 miles) of their nests. But the burgeoning capacities of the silicon chip soon eclipsed such primitive systems and the era of the penguin gadget bloomed.

Today, we have devices that record the exact three-dimensional trajectory of penguins underwater, calculating their position many times per second. This unit details every twist and turn, every single flipper-beat, and even helps us determine the energetic cost of activities, such as the radical manoeuvres penguins undertake to avoid being caught by predators or their sometimes spectacular exit from the water. The most

COURTESY RORY WILSON

sophisticated version of this tag, which has a minute beak sensor incorporated, even tells us when penguins catch their prey, how big they are and how many are swallowed in the depths of the ocean. The increasing number of highly sophisticated tags has also revealed much about penguin physiology, showing how penguin heartbeat frequency slows during diving, but speeds up again during rest periods at the surface while birds work to pump oxygen back into their bodies. This same technology has also shown that the longest, deepest dives may be accompanied by deep body cooling, which is thought to help birds stay longer underwater by slowing down the tissue need for oxygen: Emperor penguins have been recorded as diving in excess of 20 minutes, nothing short of incredible for a bird, and these mechanisms may help explain how this is possible.

Many penguin tags are retrieved after single, short foraging trips but others may be left to store their information for months, to be retrieved when the penguins return to the colonies after, for example, a winter at sea. Other systems beam information up to satellites to be relayed back to base stations as far away as France almost in real time.

However the tags may function, data on what penguins get up to at sea are always fascinating. Penguin researchers have discovered, for example, that King penguins may travel up to 600 km (375 miles) away from their colonies to find food for their chicks, that both Emperors and Kings can feed in the black depths exceeding 300 m (985 ft) — though we are still not quite sure how — and that Gentoo penguins can exceed speeds of 22 km/h (13.5 mph) while Emperors can top 30 km/h (18.5 mph). We now know that all penguins subscribe to three types of dive profile: general commuting dives, travelling in a fairly straight line only a few metres beneath the surface; 'bounce dives' when inspecting an area for prey, descending steeply and returning straight back to the surface; and hunting dives, descending to a predetermined depth before swimming along horizontally, searching for food.

Most penguin species appear to snatch their prey (normally quickly swallowed underwater) from below, and two explanations have been put forward for this. One theory holds that prey seen from below are easiest

to detect as silhouettes against the bright backdrop of the water surface. The other says that in pursuing prey from below, penguins make use of their natural buoyancy — stemming from the air in their lungs, air sacs and trapped in their feathers — to help accelerate with the least amount of effort. In fact, studies have shown that Magellanic penguins, for example, often catch fish in a rush that involves no flipper beating at all, the reverse marine equivalent of a falcon stooping onto a pigeon.

Penguins are indeed masters at controlling their buoyancy, which is important for best management of energy expenditure. They do so by controlling the amount of air that they inhale prior to a dive. The majority of a penguin's buoyancy is derived from the air in its respiratory spaces, although the volume decreases with increasing pressure at depth. In an ideal world, penguins should hunt at depths where they are neutrally buoyant, a feat they achieve by inhaling just the right amount of air so that with air compression the body attains neutral buoyancy upon reaching the target depth.

But penguins have revealed one more trick. To forage efficiently they have to maximise the time they can spend underwater and minimise wasted time at the surface replenishing oxygen supplies. Long dives require disproportionately long recovery periods so, all things being equal, birds should go for short dives and very short surface intervals. This is fine while in search mode, but breaks down once prey is encountered; nothing worse than having insufficient oxygen left to stay down and feed. They solve the problem elegantly. Since penguins mostly hunt schooling species, they usually make repeat feeding dives and 'count' how many prey they can take per dive, then replenish at least enough oxygen to allow them to catch that number, and a couple more, the next time they go down. By immediately changing dive tactics in response to the particular situation, they can maximise their efficiency at all times. Standing on my island in 1980 watching tiny penguin heads vanish into the surf, I never would have dreamt that these extraordinary birds were party to such sophistication — or that we would ever be able to find out! 🐧

ABOVE Once the Southern rockhoppers of the Falkland Islands have completed their annual moult at the onset of winter, they depart for the open ocean for four or five months. Using sophisticated satellite tracking devices during that period, Falklands Conservation has recorded individuals travelling around Cape Horn and into the Pacific, up to 1340 km (832 miles) from their home colony, tracking 4500 km (2796 miles) in the interim. In contrast, Galapagos penguins (BELOW LEFT) are the most sedentary, with dive recorders indicating most of their feeding is done within 200 m (656 ft) of shore and 6 m (20 ft) of the surface.

Marking Penguins: Minimising Research Impacts

PHOTO COURTESY L. KERNALEGUEN

Yvon Le Maho

In early December 1971, after a rude, seasickness-blighted crossing of the Southern Ocean on the Danish ship *Thala Dan*, I was fascinated by the spectacle of Adélie penguins standing and running on ice-floes under a bright sun on our final approach — my first glimpse of Antarctica.

A few days later, my colleague René Groscolas was introducing me to a study colony of those middle-sized penguins located on Petrel Island, where the French Dumont d'Urville Research Station is established in East Antarctica. The numbers on the flipper-bands of all adult breeders needed to be read, and their fledglings banded. The main objective was to obtain data on population dynamics as an indication of the effects of climate and marine resources.

This particular colony had been selected because of its location in a sloping canyon, which made it easy to capture the birds whose only chances of escape were its top and bottom ends. But I will never forget their terrified looks upon our arrival. Also, in contrast to the other Adélie penguin nesting areas, though still showing abundant signs of long occupancy, many parts of this colony had apparently been deserted in recent years. Clearly, the impact of human disturbance resulting from the monitoring

programme was too much for them to cope with. (A few years later, Pierre Jouventin, then in charge of the Antarctic ecological programme, took the right decision to abandon the canyon study site.)

Compared to other birds, which were, and still are, usually marked with leg-bands, penguins have normally been banded at the base of a flipper because their leg anatomy does not allow the fitting of a ring. Flipper-bands at the time were also considered a key advantage because the large written numbers were visible at a distance, avoiding the need to recapture the birds for their identification.

But already in the 1970s, the first concerns about flipper-bands came from observations on captive penguins in zoological gardens. Severe wounds were noted, particularly when flipper tissues were swelling during moult. Rory Wilson and his colleagues were the first in the 1990s to suspect that flipper-bands might also have a serious impact by increasing hydrodynamic drag when the birds are foraging at sea. Using a water tunnel which allowed Adélie penguins to surface and breathe only at the two extremities, they found that oxygen consumption was increased by as much as about 20% when comparing birds with and without flipper-bands. On my invitation, Rory's team brought a larger tunnel to our French research base on Possession Island in the subantarctic Crozet Archipelago, where King penguins were found to have a similar increase in oxygen consumption as Adélies after being banded.

As a precautionary measure, towards the end of the 1980s some major research teams had completely abandoned flipper-banding. But many others were still engaged in large-scale banding schemes. In the meantime, I was searching for a way to automatically monitor penguins in their colonies. In 1990, by chance I read about a new method for tagging animals. Evidently, this would enable the electronic identification of individuals of great value, such as rare species in zoos. The next day, I asked Jean-Paul Gendner, the engineer in charge of electronics in our laboratory, to investigate. He discovered that Texas Instruments Netherlands had developed the Radio Frequency IDentification (RFID) system TIRIS™ (Texas Instruments Registration and Identification System) for the

individual identification of livestock, such as cattle and pigs. The RFID tags were so small, weighing only 0.8 g (0.03 oz) that they could easily be implanted hypodermically. More recently, the microchip devices used in RFID have been referred to as PIT (Passive Integrated Transponder) tags.

With a 'partnership before commercialisation' with Texas Instruments, since TIRIS was still undergoing trials, we subsequently pioneered the RFID tagging of wild animals, deploying it on King penguins during the 1990–91 subantarctic summer at Possession Island. But RFID has a major limitation: the reading distance of the PIT tags is short, with a maximum range of about 0.6 m (2 ft) for TIRIS. Its great advantage is, however, that it allows the identification of animals which, after their initial capture for tagging, may be left undisturbed for the rest of their lives. Since the tags are implanted beneath the penguins' skin, there is no increased drag when they forage at sea, nor any danger of injury through chafing or partial detachment. PIT-tagged penguins can therefore reliably be used as 'control' animals.

Moreover, we used underground antennas along their natural pathways, which further avoids disturbance. With this in place, long-term monitoring together with my main co-workers, Michel Gauthier-Clerc and Céline Le Bohec, revealed striking conclusions. Over a 10-year period, the breeding success of a sample group of 50 PIT-tagged, flipper-banded King penguins was reduced by 40% compared to 50 PIT-tagged unbanded birds. While this may represent a fairly small sample size, at the time I did not want to flipper-band more than 50 birds as I already suspected that this would be detrimental. Furthermore, as the main subject of Claire Saraux's PhD, we found that the adult survival over the same timespan was 16% lower for the flipper-banded birds than for those without bands. David Ainley and his colleagues likewise determined that flipper-bands reduce both breeding success and survival of Adélie penguins at their own study site in the Ross Sea.

Clearly, the flipper-bands are negatively affecting the birds' swimming efficiency at sea, since both studies — on King and Adélie penguins — independently revealed that the duration of their foraging trips was also markedly longer. Additionally, we demonstrated with the Kings that in the survivors this remains so even after 10 years, proving that the birds are unable to adapt to the increased burden imposed by the bands, contrary to earlier assumptions.

Also, based on their return rate — usually after two to three years — the survival of 2500 non-flipper-banded King penguin chicks we have PIT-tagged, over 10 years was about 70–85%. This finding is significant compared to survival figures derived from flipper-banded chicks available in the literature, varying between just 6 and 45%. Moreover, we found that the responses of flipper-banded King penguins to climate variability also differed from those of non-banded birds, therefore indicating that relying on data from flipper-banded birds introduces a serious bias in determining and predicting the impact of climate changes on penguins.

Combining RFID with an electronic scale also allows automatic weighing. Already in our 1990–91 pioneering set-up on Possession Island, we could calculate the weight gain of a King penguin returning from foraging at sea, an indicator of marine resources, or its weight loss while in the colony, where it only relies on its body fuels. Today, the combination of scales and RFID systems is being used by scientists from different nations to monitor various penguin colonies around the world: Little blue penguins at Phillip Island, Australia; Adélies at Cape Crozier and Terra Nova Bay in the Ross Sea, as well as near Mawson and Davis stations; and Macaroni penguins at South Georgia's Bird Island. We have even installed such an automatic system, avoiding disturbance at last, for about 250 pairs of Adélie penguins today located in the once deserted canyon near Dumont d'Urville Station.

ABOVE Exquisite streamlining, seen here in porpoising Gentoos, may be affected by the smallest turbulence. OPPOSITE TOP A small metal band fitted to a King penguin's flipper appears minor, but may have a major impact on the bird's life, requiring more energy expenditure with resultant lower life expectancy and breeding success.

BELOW High speed underwater manoeuvres are key to a King penguin's survival. OPPOSITE BOTTOM An independent study in the Ross Sea also found lower breeding success in flipper-banded Adélie penguins than unbanded birds.

The Crested Penguin Egg-size Conundrum

Kyle W. Morrison

A Canadian national, Kyle Morrison became fascinated by seabirds while studying how early conditions and climate events impact survival rates of auks — the northern counterparts of penguins. His PhD at Massey University, in collaboration with New Zealand's National Institute of Water & Atmospheric Research (NIWA), has taken him to Campbell Island where his research considers how foraging behaviour, diet, predation and demography interact to determine the dynamics of the island's much-reduced Southern rockhopper penguin population.

Ecology Group, Institute of Natural Resources Massey University, Private Bag 11-222, Palmerston North, 4474, New Zealand k.w.morrison@massey.ac.nz

Synopsis: The remarkable size difference between first- and second-laid eggs in all crested penguins, together with a marked difference in incubation time and the near-certain loss of the first egg or chick, has led to many intriguing theories, with the answers probably lying in the distant evolutionary past.

Would you invest in something with next to no hope of receiving a return? The seven species of crested penguins (genus *Eudyptes*) do just that: they lay two eggs, but almost never rear two chicks. Equally perplexing, crested penguins lay extremely dimorphic eggs, where the first-laid egg is much smaller than the second. Depending on the species, this 'A-egg' averages only 54–85% the mass of the second-laid 'B-egg'. Erect-crested penguins lay the most dimorphic eggs, followed by Royal and Macaroni penguins; these three species almost invariably lose their A-egg early in incubation. Southern rockhopper, Northern rockhopper, Snares and Fiordland penguins may hatch both eggs, but rarely succeed in rearing both chicks (< 3% of clutches). Even when both eggs hatch, the A-chick usually starves to death within a week, non-aggressively out-competed for food by the larger B-chick. The B-chick is bigger not only because it emerges from a larger egg, but also because, surprisingly, it usually hatches a day before the A-chick — despite the B-egg being laid 4–5 days after the A-egg!

Crested penguins are unique among birds that lay small clutches in several ways: for the greatest size difference between eggs, for laying their largest egg last, and for having their last-laid egg hatch first. These traits raise many questions that have intrigued scientists for decades, but have defied a comprehensive explanation.

The chance to investigate these mysteries adds eagerness to my descent of the steep slope to Penguin Bay's Southern rockhopper colony on New Zealand's subantarctic Campbell Island. It is early November, the time females begin to lay their first eggs. In the four- to five-day interval between the appearance of the A- and B-egg, I observe pairs primarily taking turns standing over, or next to, their A-egg, rather than incubating it properly. Some pairs treat this first egg as if it were a rock in the nest, rather than the viable potential offspring that it is. I see a few A-eggs almost completely buried by pebbles, others stepped on and broken, some stolen by skuas, and a number simply lying outside the nest, sometimes within reach of the parents, but neglected.

Predictably, soon after laying commences is one of the peak times for A-egg failure in crested penguins, especially in the three species that lay the most dimorphic eggs. In Royal penguins, St Clair et al (1995) noted that A-egg loss is frequently associated with female nest-building behaviour during the 24 hours before the B-egg is laid. The authors interpret this as deliberate egg ejection and evidence of maternal infanticide, and suggest the same is likely true of Macaroni and Erect-crested penguins. In contrast, Davis and Renner (2003) propose A-egg loss in Erect-crested penguins to be the result of neglect and the mechanical difficulty of incubating an A-egg that is about half the size of the B-egg.

RIGHT The greatest egg-size differences of all crested penguins is seen in the Erect-crested, whose A-egg may be nearly half the mass of the B-egg (far right, in hand), compared to the Southern rockhopper clutch.

The four other crested penguin species often incubate both eggs until hatching, with a peak in A-egg failure occurring after the B-egg hatches, but before the A-egg does. During this period of about a day, the movements of the chick and the brooding parent frequently result in the ejection of the A-egg from the nest. Such eggs are often already pipped, or cracked, and beginning to hatch. Incredibly, even the chick's peeping from inside the egg usually fails to stimulate an egg-retrieval response from its parents.

A comprehensive explanation of the crested penguin conundrum must address three key questions: (1) why is it so rare for crested penguins to rear two chicks; (2) why do crested penguins lay a two-egg clutch; and (3) why is the A-egg so much smaller than the B-egg?

Our first question requires us to consider the selective pressures that result in the almost invariable loss of an egg or chick, rather than the direct causes of these failures. A trait shared by the majority of a group of related species was very likely inherited from their most recent common ancestor, rather than evolving independently in each species. All seven crested penguins lay dimorphic eggs and make long-distance foraging trips offshore while breeding, so these characteristics were probably present in the ancestral crested penguin. An offshore-foraging strategy places greater energetic demands on parents than those faced by inshore-foraging penguin species (Davis and Renner 2003), perhaps explaining why it is so rare for crested penguins to successfully rear two chicks. In contrast, the sister-genus of *Eudyptes* is the non-crested *Megadyptes*, represented solely by the Yellow-eyed penguin, an inshore-foraging species that lays two eggs of equal size and frequently raises two chicks.

Turning to question (2) Why lay two eggs if only one survives?, with the exception of Kings and Emperors that lay one egg, all penguins lay a two-egg clutch, so the latter is likely the ancestral trait of all penguins. If the Emperor/King lineage was able to develop a one-egg clutch, then why not the ancestral crested penguin likewise? The most plausible explanation is that the B-egg had a better chance than the A-egg of producing a viable offspring. If so, this higher fitness value of the B-egg would select against elimination of the second ovulation that produces it (Williams 1980).

This explanation implies that a two-egg clutch is a maladaptive trait in crested penguins, but other adaptive hypotheses propose ways in which the A-egg confers a fitness benefit that outweighs its energetic cost of production. These include A-eggs serving to synchronise laying, increasing nest site or mate retention, acting as insurance against failure of the B-egg/chick, and allowing the potential to raise two chicks under particularly favourable conditions. However, these theories tend to fall down when one considers that the overall cost of A-egg production is not just in the energy contained in the egg, but also in the extra four to five days of fasting before the B-egg is laid, and that three crested penguin species almost invariably lose their A-egg very early in incubation.

Furthermore, these hypotheses do not address question (3), Why is the A-egg so much smaller?

If their offshore-foraging strategy makes crested penguins generally incapable of rearing two chicks, and a higher value of the second-laid egg selects against the evolution of a one-egg clutch, it follows that females should preferentially invest in their B-egg. The essential question then becomes, what caused the A-egg/chick to be less successful? The most widely cited explanation is based on the reasonable assumption that, because the Yellow-eyed penguin and the crested penguins so far investigated share the delayed development of a complete brood patch until B-egg laying (and all delay full incubation until then), this was an attribute inherited from their common ancestor. This incubation delay and associated egg-chilling presumably increased the risk of A-egg failure through displacement, predation and a longer incubation period, leading to preferential investment in B-egg formation by the ancestral crested penguin (St Clair 1992). If the ancestral Yellow-eyed penguin had the relatively more solitary, sheltered nests of its descendant, risk of failure would be more similar between A- and B-eggs, removing the selective pressure to produce a big B-egg and small A-egg.

Crested penguins share the unfortunate distinction that each species is classified as 'Vulnerable' to extinction, or worse (IUCN 2012). Although scientists may never agree how the crested penguin egg-size conundrum originated millions of years ago, they would likely accept that this reproductive strategy appears ill-suited to recent ecological conditions. Future studies combining ecological information with new techniques in reproductive and developmental physiology promise fascinating new insights into the mysteries of this complex and confounding topic. 🐧

OPPOSITE Southern rockhoppers stand at their nest sites ignoring their first-laid eggs. These A-eggs remain neglected until the B-eggs are laid, when both will be incubated.
ABOVE After the successful hatching of the B-egg, a brooding Southern rockhopper on Campbell Island neglects its A-egg even in the process of hatching, remaining indifferent to the chick peeping within.
BELOW Erect-crested penguins raise only one chick, usually losing their A-egg before incubation begins, Antipodes Island.

Emperor Youngsters on the Move

Barbara Wienecke

After spending most of her early childhood in Namibia, where she was born, Dr Barbara Wienecke went to high school in Germany before moving to Australia to pursue an education in biology. She obtained her PhD in seabird ecology from Murdoch University in Western Australia. Now living in Tasmania, she has been working with the Australian Antarctic Division since 1993, starting off with a winter spent with the Emperor penguins near Mawson Station in East Antarctica. Her work, both in Antarctica and the Subantarctic, involves tracking studies of Emperor and other penguins, plus some flying seabird species, along with long-term population monitoring of key Emperor penguin breeding colonies.

Australian Antarctic Division Channel Highway Kingston, Tasmania 7050 Australia Barbara.Wienecke@aad.gov.au

Synopsis: Long-range satellite tracking of young Emperor penguins in East Antarctica reveals some extraordinary facts on their diving capabilities and oceanic ranging.

When the sun is again high in the skies above the frozen landscape of Antarctica and the recently returned Adélie penguins and petrels are busy with their breeding activities, Emperor penguins are at the end of their reproductive season. Life in their ice-bound colonies begins to change rapidly. By mid-December, the chicks have grown to about 80% of their parents' size. On a wind-still sunny day, the temperatures can reach 4 or 5°C, far too hot for Emperor penguins to feel comfortable. Even at 0°C the penguins pant or lay spread-eagled on the ice. They often wander around looking for fresh snow, which they eat to remain hydrated. The chicks are losing their soft grey down and are growing waterproof feathers. Soon they will start fending for themselves out in the ocean. The moult is energetically expensive, and the chicks' metabolism is working fast to produce thousands of feathers. I wonder whether growing feathers itches; the penguins often look so uncomfortable, especially on a warm day.

By late December, the number of adults in the colony is greatly reduced; many of them have already left their chicks. Most chicks have not yet realised that their parents will not come back with another stomachful of food. They are begging to be fed by any returning adult. Eventually the young penguins comprehend that the adults are heading away from the colony, so they start wandering in the same direction.

For the fledglings, it is their maiden voyage, their first ever contact with the sea. For their parents, this trip is crucial, as they will need to forage intensively to gain sufficient body reserves to survive their annual moult. It is at this time that I wanted to be at the colony to find myself some 'volunteers' to carry satellite trackers for a few weeks or months. This would address some key questions I was trying to answer: Do adults and fledglings utilise the

same feeding areas? Do they remain in the pack-ice zone or will they forage in open oceanic waters? Will the fledglings return to the continent in winter?

Over a number of years, I deployed over 50 satellite trackers on both adult and fledgling Emperor penguins in East Antarctica at three colonies accessible from the Australian stations. The Auster colony (east of Mawson) comprises about 11 000 breeding pairs and, like most Emperor penguin colonies, is located on the land-fast sea-ice among grounded icebergs. In contrast, the colony at Taylor Glacier (west of Mawson) has only about 3000 breeding pairs, and is unusual because it is one of only three known land-based breeding colonies. Since 2009, Emperor penguin fledglings have also been tagged at Amanda Bay (south west of Davis) in Prydz Bay, home to approximately 10 000 breeding pairs. Seven fledglings were fitted with satellite trackers at Taylor Glacier in 1996. Ten years later, in 2006, I repeated the experiment at Auster and deployed 10 tags.

Deploying tags on Emperor penguins is not an easy task. The birds are big and strong and even the fledglings know how to fight. The worst thing though is that as much as I love being near them, I know how stressed they become when handled, so the work is done as quickly and as efficiently as possible. Before attaching the devices we weigh each fledgling to ensure it is big enough. They need to weigh at least

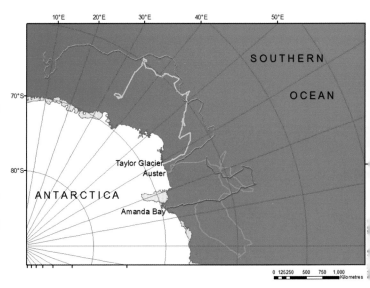

13 kg (28.5 lb). The satellite trackers are only 92 g (3.2 oz), powered by two AA batteries, and to conserve power are programmed to transmit data for only four hours in 48.

Once all the trackers are deployed the excitement begins. Where will the penguins go? How far will they travel? The results were rather surprising.

Upon departing the colony, the Auster fledglings had to cross nearly 50 km (31 miles) of fast-ice, and would not be able to feed for several days, until they reached open water. Most of them set off walking north towards the ice edge, but one headed in the opposite direction heading for the continent for nearly two days before realising its mistake and turning around. Once they reached the edge of the fast-ice the young Emperors had a wide band of pack-ice in front of them. It was remarkable to see how they moved through it, travelling directly north towards the deep waters of the Southern Ocean.

Within days, they had left the pack-ice behind, and after about a fortnight at sea they began to disperse. Most fledglings ventured to areas as far north as ~54°S. In a straight line, that is 1500 km (932 miles) north of their home colonies. Of course, penguins never travel in such straight lines so the actual distances they covered while meandering around the Southern Ocean were far greater. After six and a half months at sea, some had traced nearly 6600 km (4100 miles) on the map (see diagram below left). Collectively the 17 youngsters ranged across nearly a quarter of the Southern Ocean (7°E to 93°E) in their first half year at sea — over 2500 km (1550 miles) from their natal colony (Wienecke et al 2010).

These results were so exciting that I decided to see how typical such travels might be. In 2009 and 2010, I went to Davis Station to conduct a similar study at Amanda Bay. This bay is in the south-eastern corner of Prydz Bay which, from an oceanographic perspective, is very different compared to the Mawson Coast. The fast-ice edge is much closer to Amanda Bay than it is at the Mawson Coast, and Prydz Bay is a rich feeding area for whales and seals, which presumably serves

penguins too. Fascinatingly, the youngsters again engaged in a strong northwards movement. But overall they travelled slower than their colleagues in the west. Some of them again spent an extended time in the middle of the Southern Ocean far away from land or ice. Similar northward movements were recorded among Emperor fledglings in the Ross Sea (Kooyman and Ponganis 2008). A brief look at ocean currents in relation to the penguins' travels revealed that far from passively riding the currents as might be expected, they regularly swim against them. Clearly they are making their own decisions about where to go once they are out there in the big, deep, blue ocean. Yet the big question remains unresolved: what drives them to go so far north?

The satellite tags also transmit dive information. The diving capacity of young Emperor penguins develops quite rapidly. Both maximal depths and dive duration increase significantly in their first few months at sea. On their first day at sea, the fledglings dived regularly to 40 m (131 ft), submerging for about two minutes. Three months later, some of them reached depths of over 300 m (984 ft), which for comparison is approximately the maximum dive depths of adult King Penguins. Mature Emperor penguins go even deeper, the deepest recorded in East Antarctica attaining a staggering 564 m (1850 ft) (Wienecke et al 2007). However, routine dives usually do not exceed about 100 to 120 m (328–394 ft). It may take as much as two years for the juvenile Emperors to fully develop their diving capacity. What probably limits their ability to attain depth initially is the myoglobin (oxygen-binding protein) concentration in their muscles. Fledglings have only one-third the concentration of myoglobin found in adults, although other proteins necessary for deep diving are virtually at the adult level when they first go to sea (Ponganis et al 1999). Nevertheless, even from the start, young Emperor penguins are already exquisite divers; by the time they mature, their prowess is unrivalled among all other known seabirds — and quite a few marine mammals too. ⬆

ABOVE A lone adult in normal summer habitat in extensive areas of drifting, open pack ice, East Antarctica.
BELOW The author at her study site, surveying the 10 000-strong Amanda Bay colony in East Antarctica.
OPPOSITE TOP Abandoned on the ice by their parents, fledglings head for the sea, Cape Darnley, East Antarctica.
FAR LEFT A fledgling (left) departs partly down-clad and weighing only about half the adult weight; it will return fully grown (right) the following summer for its first moult.

Little Blues: Smallest Penguins Face Big Challenges

André Chiaradia

Originally from Brazil, Associate Professor André Chiaradia is an oceanographer involved with seabird ecology as part of a small but very active research group at Phillip Island Nature Parks, Australia. He began studying albatrosses and petrels on fishing boats off the Brazilian coast but currently works on the Little blue penguin, focussing on fine-scale responses in a top predator to natural changes in the marine system and predictions for future environmental change. He is enthusiastic about open source journals (on the editorial board of three) and the revolution of R free software in science.

achiaradia@penguin.org.au
www.penguin.org.au

ABOVE RIGHT Rotund, well-fed chicks explore outside their nest burrow, Bruny Island, Tasmania.
BELOW Pair mating in the dead of night, Bruny Island, Tasmania.
BELOW RIGHT The shiny blue coloration of this penguin's feathers derives from light refraction through micro-fibre structures.

Synopsis: Detailed, long-term analysis reveals how a local penguin population recovered and stabilized by shifting its prey base in the aftermath of the near-total collapse of sardine banks along the south Australian coast.

In 1995, along the coast of Australia and New Zealand, a massive mortality decimated the population of sardines *Sardinops sagax*. Daily news showed tonnes of dead fish washed up on beaches. It was the largest single fish species mortality ever recorded in any marine system (Jones et al 1997), and Little blue penguins were soon caught up in its aftermath. At the time, I had just started my PhD on the foraging ecology of these penguins. Suddenly, sardines in their diet, or lack thereof, became a vital part of my work.

The epicentre of this mortality was around marine fish farms, where imported sardines had been used to fatten captive tuna. From there, the sardine mortality spread an impressive 35 km (22 miles) a day, decimating some 70% of the population over 5000 km (3100 miles) of the Australian coast, seriously depleting the southern Australian stocks in a very short time (Gaughan, Murray and colleagues, several papers). A herpes virus, common to imported sardines but not to the local population, was found in the dead fish (Whittington et al 2008). A vital component of the marine food web, sardines featured in the diet of several top predators, including penguins, and was one of the most abundant commercial catches in southern Australia's coastal fishery.

What happened next was entirely predictable. The commercial fishery collapsed spectacularly. In the State of Victoria alone, catches went from over 2300 tonnes in 1995 to less than 100 tonnes after that (Fisheries Victoria Information Bulletin). These drastically reduced catch levels continue today.

Little Blue penguins felt the effects too. In 1997 they suffered their lowest ever breeding success since records began in 1968. Sardines, once a major prey, virtually disappeared from their diet. In a perverse way, this mortality provided a natural experiment in which the trophic response of such a top predator to changes in its marine ecosystem could be examined. To monitor penguin diet without too much disturbance, we incorporated stable isotope analysis of the blood and DNA analysis of scats to determine at what level in the food chain the penguins were feeding (Cavallo et al 2018, 2019; Chiaradia et al 2010, 2016)

Ten years after the 1995 mortality, we found that penguins were targeting prey higher in the trophic pyramid which, in theory at least, means a lesser abundance of food. Surprisingly, however, all indices of penguin breeding success were not as bad as expected. Sensitive measurements of parental foraging effort, like chick peak mass and linear growth rates, increased steadily, with their breeding success generally becoming more stable than in the 1980s and early 1990s.

The surprising resilience of this species is thanks to remarkable individual qualities. Little blues have revealed some cunning strategies. They can exhibit a good travel plan in their foraging decisions, as demonstrated in various studies using bio-loggers, automated weighing stations and fractal movement analysis. Some individual penguins work harder than their partners to feed their chicks, the over-achiever parents making more foraging trips than their mates. This situation is the norm (72% of cases), rather than the expected equal parenting (Saraux et al 2011a).

Individuals can also alternate between two consecutive long foraging trips and several shorter ones throughout the chick-rearing period, a strategy rarely observed for inshore seabirds (Saraux et al 2011b). Short trips allow for regular food provisioning of chicks (high feeding frequency and larger meals), whereas longer trips are triggered by a parent's low body mass and, therefore, the need to replenish its own energy reserves. Also, the groups of penguins crossing the beaches are not formed by accident: individuals seem to choose

their travelling partners (Daniel et al 2007), while some even take advantage of artificial features, like ship channels, to aid in their foraging (Preston et al 2010).

Age too seems to play a crucial role, as middle-aged penguins are better breeders (Nisbet and Dann 2009), employ more effective foraging strategies (Zimmer et al 2011) and feed in different locations (Pelletier et al 2014, Sanchez et al 2018). Finding out about individual qualities may sound trivial, but overall it can make a big difference. It is vital to understand these variations before making connections with environmental changes, since individuals respond differently to the same environmental pressures.

Given their broad geographic range, Little blue penguins are not listed as endangered in the IUCN Red List (see Part III of this book). However, many colonies are severely threatened by introduced predators, coastal development, oil spills and gillnet fishing in Australia and New Zealand. Remarkably, these pressures can be locally offset by a wide range of conservation efforts. For example, close to my home at Phillip Island, Australia, in recent decades roads were closed, introduced predators like red foxes systematically eliminated, and entire housing estates have been bought out and demolished by the State Government. Restoring natural conditions on land helped reverse the bleak prediction that this penguin population, now numbering around 32,000 penguins (Sutherland and Dann 2014), would vanish by 2000 (Dann 1992). Other areas have not been so lucky, such as Marion Bay in Tasmania, where a nesting colony I visited as a potential study site in 1994 no longer exists (Stevenson and Woehler 2007).

By breeding on land and foraging at sea, penguins often face problems in both environments. While pressures ashore stand out easily in our minds, troubles at sea, where they spend about 80% of their lives, are out of sight and therefore much more difficult to quantify. In studies with excellent animal diving scientists Yan Ropert-Coudert and Akiko Kato, we found that Little blue penguins hunt more successfully in shallow waters, probably because the seafloor's foraging zone is physically limited by thermoclines (steep and sudden temperature gradients) in the water column. While warm-blooded penguins can travel through different temperature layers, their cold-blooded prey may be slowed down or cannot cross

these thermal barriers altogether, so it becomes an easy catch (Meyer et al 2020, Pelletier et al 2012). These are local features that can be affected by global events. For example, thermal layers can get mixed after storms — which are predicted to increase in El Niño years — making it harder for Little blue penguins to find prey dispersed in the water column.

We should not react complacently to the remarkably adaptable response of Little blue penguins to the virtual disappearance of a major prey species. Since the sardine mortality, the penguin food web has become simpler and less flexible, making them more vulnerable to further changes in their marine system. Other risk factors are also emerging within the distribution range of Little blues. In this corner of the planet, we are already experiencing some rapid climatic changes (see Voice et al 2006). The catastrophic breeding season the penguins experienced in 1997 was blamed on the sardine die-off, but it also coincided with a strong El Niño year. The synergy between the absence of a major prey coupled with an El Niño event may have created a 'perfect storm' to collapse the penguin's breeding abilities that year. So the world's smallest penguins are also facing some of the biggest challenges of our planet. ▲

ABOVE Tiny penguins land in splattering waves on the boulder shores of Phillip Island, Australia. ABOVE LEFT The uniquely pale White-flippered subspecies finds shelter in the no-fishing Pohatu Marine Reserve, Banks Peninsula, New Zealand. BELOW Guaranteed penguin viewing from grandstand seating at the Phillip Island Nature Parks attracts tens of thousands of visitors from around the world, who, by watching the penguins throng ashore undisturbed every evening, have financed their comeback through active conservation management.

Galapagos Penguins: An Uncertain Future

Hernán Vargas

Dr Hernán Vargas is Director of The Peregrine Foundation's Neotropical Science and Student Education Program. He obtained his first diploma from the Universidad Católica del Ecuador in 1989, followed by an MSc at Boise State University, USA in 1995. After six years as resident ornithologist at the Charles Darwin Research Station (1995–2001), a research student position at Oxford University, UK (2002–2005) culminated in a PhD in conservation biology in 2006.

The Peregrine Fund (TPF)
Casilla 17-17-1044
Quito, Ecuador
hvargas@peregrinefund.org

ABOVE RIGHT Nesting in lava cave, Mariela Islets.
BELOW Sharing volcanic habitat with other tropical species, Isabela Island.

Synopsis: With a total population fluctuating between some 800 and 3400 individuals over the past 40 years, intense scrutiny reveals a bevy of threats for the only equatorial penguin.

The Galapagos penguin is arguably one of the most unusual members of its family, being the only species living in equatorial waters. But even though I grew up in the Galápagos Islands, our respective home ranges didn't overlap; its habitat is restricted to very specific areas flushed by cold, nutrient-rich upwelling currents, mostly along the western edge of the archipelago, whereas mine was on centrally located Santa Cruz Island, where penguins almost never stray.

This changed in 1978 when, as a high school student, I volunteered as a field assistant with biologists from the Charles Darwin Research Station. My assigned task — to collect and analyse feral dog faeces for dietary composition — documented for the first time, one of a fleet of human-induced threats to this critically endangered species.

Dogs were eradicated by the Galápagos National Park not long after, but my work has brought me back to the plight of Galapagos penguins ever since. To date much has been learned through a battery of complementary studies, all of which have cast further light on the Galapagos penguin's deepening troubles, together with efforts to mitigate human impact where possible.

A major threat appears directly linked to climate change, which historically has driven increasingly frequent, intense and prolonged El Niño events. Due to their coastal foraging habits and very restricted range (Steinfurth et al 2007; Vargas 2006), when high sea temperatures cause small schooling fish to disperse, penguins are left with no food and no place to go. The notoriously strong El Niño events of 1982–83 and 1997–98 caused mortalities of up to 77% of the total Galapagos penguin population (Valle and Coulter 1987, Boersma 1998, Vargas et al 2006, Vargas et al 2007). Several El Niño-free years after 1998 allowed a partial recovery, yet by 2009, when the last complete census was conducted (Jiménez-Uzcátegui and Devineau 2009), the total number of individuals counted was 1042, a sobering 40% below the pre-El Niño levels (see diagram below).

Unfortunately, with widespread climate change derived from global human activity, the best conservation strategies are limited to alleviating the effects of other, more manageable threats. One recent initiative, spearheaded by Dee Boersma in collaboration with the Galápagos National Park, involves the construction of shady lava shelters to increase the availability of potential nest sites. Based on the theory that a dearth in quality natural nesting caves may act as a limiting factor on breeding success during favourable La Niña conditions — when abundant food supplies

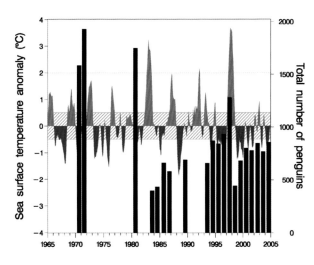

DIAGRAM H. VARGAS (2006)

could allow each pair to fledge two or three broods within a calendar year — it is hoped that this will serve as a hedge to offset catastrophic famines when El Niño strikes again (Boersma pers comm).

Another area of great concern is the effect, real and potential, of pathogens and parasites brought to the islands by human activities. Through collaboration with Professor Patricia Parker and her team from the University of Saint Louis, we have made great progress in assessing the situation as well as determining normal health parameters of individual penguins (Vargas 2009, Parker 2009 in De Roy 2009). An alarm was raised when a protozoan parasite of the genus *Plasmodium* was detected in penguin blood samples taken between 2003 and 2006. This parasite closely resembles *Plasmodium elongatum* which, together with *Plasmodium relictum*, are highly pathogenic parasites elsewhere in the world, responsible for deadly outbreaks of avian malaria in captive Humboldt penguins (Huff and Shiroshi 1962), the Galapagos penguin's closest relative (Bollmer et al. 2007). Neither of these two species has been found in the Galápagos (Levin et al 2009). This finding raised questions which we are still trying to answer: (1) what is the exact species of this parasite? (2) How closely is it related to the other two species? (3) What are the likely vectors and reservoirs in the Galápagos environment? And 4) What are the conditions for this parasite to become pathogenic? So far we have not succeeded in identifying the *Plasmodium* vector, although the mosquito *Culex quinquefasciatus*, introduced to the Galápagos in the early 1980s (Whiteman et al 2005, Parker et al 2006) is the most likely candidate, because of its documented role in spreading avian malaria in Hawaii (van Riper et al 1986; Atkinson et al 2000).

In our attempts to elucidate the matter, we are collecting mosquitoes to assess distribution patterns as related to *Plasmodium* prevalence in penguins, mapping freshwater sources where *Culex* can breed and testing blood from other birds sharing the penguins' range to establish whether these may act as reservoirs of parasites. The good news is that no mortality has been associated with any of these parasites, and the more than 500 penguins sampled throughout this exercise were all found to be in good health. But we now hypothesise that mass penguin mortality during major El Niño events could result from synergistic effects of food stress combined with *Plasmodium* parasites (Palmer et al in preparation). Fortunately, testing for West Nile Virus and Avian influenza, both diseases of recent worldwide concern, have so far proven negative in Galapagos penguins. Long-term monitoring of penguin health will be needed to help us understand these complex dynamics.

As if these major risks were not sufficient, other problems facing the penguins include predation by introduced animals, such as feral cats and dogs which are known to kill adults, and rats that can take eggs and nestlings. Although the use of gillnets is prohibited within the Galápagos Marine Reserve, illegal fishing activities have been implicated in reports of penguins drowned in such gear, yet the extent of the problem remains impossible to gauge. Tourism also may have an impact, as yet unstudied. While penguins living in areas of heavy visitation seem to become easily oblivious to human attention — swimmers and small boats often approach within mere metres — it is the potential transportation of pathogens and their vectors into penguin habitat, as mentioned earlier, that remains the biggest danger. The high demand for fuel supplies for tour ships also carries the constant risk of lethal oil spills of disastrous proportions.

For decades, the Galápagos National Park has carried out relentless campaigns against introduced vertebrate pests, including expanding successes in ridding islands of their rat plagues. To control fishing activities, a number of no-take zones have been established along key stretches of coastline, and tourist access is strictly limited to a very small number of sites where penguins can be seen. An avian disease surveillance programme was initiated in 2001 by the Saint Louis Zoo and the University of Missouri–St Louis in cooperation with the Charles Darwin Foundation and Galápagos National Park (Parker et al 2006, Parker 2009).

The most stringent application of the existing quarantine system for Galápagos, plus substantial strengthening of many measures already in place, such as fumigation of all forms of human transport, from mainland Ecuador and between islands, will be our only hopes for stemming the tide of foreign organisms that threaten the integrity of the insular ecosystem — and this most vulnerable of penguins in particular. 🐾

THIS PAGE El Niño events and volcanic eruptions are among a host of threats for a highly sedentary species at the limit of climatic possibilities for penguins.

African Penguins:
A Troubled History

Peter Ryan

Prof. Peter Ryan is Director of the FitzPatrick Institute of African Ornithology, University of Cape Town. He studies seabird ecology and conservation, bird evolution, marine litter and island conservation and management. In addition to more than 400 scientific papers, Peter has written 13 books on birds and Southern Ocean islands.

FitzPatrick Institute of African Ornithology
University of Cape Town
Rondebosch 7701
South Africa
www.fitzpatrick.uct.ac.za

Synopsis: The African penguin was the first to become known to modern civilisation, but 500 years of history retrace the unhappy outcome of this relationship, while pointing towards some solutions.

The African penguin — also known as the Jackass or Black-footed penguin — has had a long and not very happy relationship with humans. Long before being sighted by Vasco da Gama's crew in 1497, they were eaten by Strandloper, hunter–gatherer communities. But the arrival of Europeans with ships able to reach their breeding islands started five centuries of human impacts that have reduced the species to only a few percent of its original population. Initially, the main threats were direct exploitation and degradation of breeding habitat through guano collecting and introduction of alien species such as rabbits. Still numbering in their millions at the start of the twentieth century, collection of more than 100000 eggs each year reduced the population to fewer than 150000 pairs by the 1950s. Subsequent protection hasn't improved the species' fate. Today it suffers from oiling and food shortages following fishery collapses in Namibia and a shift in fish stocks off South Africa. Currently fewer than 20000 pairs survive, 80% in South Africa.

My first hands-on experience with penguins came as an undergraduate student in the early 1980s when I was privileged to work as a field assistant to Rory Wilson, then a doctoral student studying African penguin foraging ecology. Ever the innovator, Rory developed speed and depth gauges that used radioactive beads to

record on X-ray film detailed profiles of the penguins' foraging trips. Although crude compared to the digital devices available today, it gave novel insights into the behaviour of penguins at sea.

Encouraged by John Cooper, I spent one vacation investigating the peculiar phenomenon of partial head moult among immature African penguins. I showed that adults discriminate against birds in juvenile plumage, so an early moult into an adult head pattern probably confers an advantage to young penguins, allowing them to join adult feeding groups. The work dovetailed with Rory's work on how the bold adult plumage of *Spheniscus* penguins (and other predators of small pelagic schooling fish) promotes the formation of 'bait balls' and thus enhances their feeding success. In the last few years, Alistair McInnes has obtained video footage from foraging penguins which confirms the importance of group foraging in this species.

The early 1980s saw the formation of three new African penguin colonies, when conditions on Dyer Island became so unfavourable that young birds sought breeding opportunities elsewhere. In 1985, one pair of penguins bred at Boulders, a sleepy seaside suburb on the southern Cape Peninsula. They were soon joined by others, and as their numbers grew, the penguins drew the ire of some residents who objected to their braying calls, pungent guano smell, and the growing numbers of tourists they attracted. There were calls for the colony to be moved to a nearby nature reserve, but the very reason for its existence at Boulders was the buffer provided by the suburb against terrestrial predators. Ultimately the only solution was improved management of the colony to limit its impact on adjacent homeowners.

The 1990s were comparatively good times for penguins around South Africa. Several years of extremely good recruitment for anchovies and sardines saw penguin populations increasing for the first time since records were kept. Their success was however tested by catastrophic oil spills. A detailed study by Anton Wolfaardt of the fate of the 10000 penguins rehabilitated after being oiled following the sinking of the *Apollo Sea* in 1994 revealed that, although most penguins survive oiling if treated quickly, their long-term reproductive fitness is impaired. The sinking of the *Treasure* in 2000 oiled 19000 birds. To keep other adults safe, a further 19500 penguins were caught on Dassen

ABOVE RIGHT Adult and chick resting outside their nesting burrow, Boulders Beach, South Africa. RIGHT Parents come ashore in the afternoon at Boulders Beach, returning to feed their chicks at the colony a short distance inland.

and Robben islands before the oil reached them, and transported 1 000 km to Port Elizabeth. Despite uncertainty as to what these penguins would do, Rob Crawford and Les Underhill put satellite tags onto three birds, so the whole world could watch online as they all swam back home. The penguins' trip home took them between 11 and 17 days, long enough for the oil clean-up to proceed.

Unfortunately, the boom years of the 1990s didn't last. From the turn of the millennium, pelagic fish stocks shifted their range from their traditional strongholds off the west coast of South Africa to the Agulhas Bank off the south coast. Environmental changes appear to have driven this shift, exacerbated by greater fishing effort off the west coast, where fishing ports and processing plants are concentrated. Penguin populations in this area crashed, while there are no suitable breeding islands along most of the south coast. Local fish shortages focussed attention once again on penguin foraging ecology. One interesting discovery was that, like many petrels, penguins respond to dimethyl sulphide (DMS), a compound that is released by phytoplankton when their cells are damaged. African penguins probably are attracted by DMS because it indicates areas where phytoplankton is being grazed by zooplankton and sardines.

In collaboration with David Grémillet from the French CNRS and honours student Samantha Petersen, we obtained the first detailed foraging tracks and dive data for African penguins using GPS and depth loggers in 2003. Post-doctoral fellow Lorien Pichegru used this technology to test whether spatial management of fishing effort can improve the fate of African penguins. Experimental fishing closures within 20 km (12 miles) of key breeding islands have been ongoing since then, with seemingly clear benefits to the penguins. Sadly, these results are still disputed by some fishery consultants, and the South African government has yet to decide on long-term restrictions on fishing close to breeding colonies.

Of course, a lack of fish doesn't only affect breeding birds. Richard Sherley showed how many juvenile penguins still disperse north up the west coast of southern Africa, to an area historically rich in prey, but which now supports few fish. Similarly, adult penguins that disperse to this area to fatten up prior to moulting show lower survival than birds from colonies on the south coast, where fish are more abundant.

The species' stronghold has shifted from the west coast of South Africa and Namibia to two island groups in Algoa Bay. However, even there they are not secure. Ship-to-ship refuelling, illegal under South African law, recently started close to St Croix Island, and coincided with an alarming fall in the numbers of breeding penguins.

As their numbers dwindle, the impacts of predation become more severe, and group-based foraging options diminish. Many breeding attempts fail when heat waves force adults to abandon their nests, or nests are flooded during storms — problems that are only likely to get worse as climate change progresses. Disease outbreaks, such as avian influenza in 2018, kill hundreds of birds that the population cannot afford to lose. One way to bolster the faltering penguin population is to establish a new penguin colony along the south coast. BirdLife South Africa has built a predator exclusion fence around a headland in De Hoop Nature Reserve, where penguins have attempted to breed in the past, and is using decoys and recordings to attract penguins to the site. Ultimately, the long-term conservation of the African Penguin will require careful management both of its breeding colonies, and its feeding areas offshore.

Magellanic Penguins: Living with People

P. Dee Boersma

Dr P. Dee Boersma began
her penguin work in the
Galapagos Islands, where
she lived for over a year
on uninhabited Fernandina
Island while working on the
Galapagos penguin. For 30
years she and her students
have studied Magellanic
penguins at Punta Tombo,
Argentina.

Wadsworth Endowed Chair
in Conservation Science
Department of Biology,
University of Washington,
Seattle, WA 98195-1800, USA
boersma@u.washington.edu

ABOVE RIGHT AND BELOW
Habituated to seeing
tourists passing by, the
endearing courtship
antics of these boldly
marked penguins make
them far more valuable
alive than dead, attracting
well over 100 000 visitors
per year to Punta
Tombo, Argentina.

Synopsis: Three decades of research at Punta
Tombo, Argentina, paints a clear picture of
this species' fluctuating fortunes along the
Patagonian coast.

My first penguin encounter in the wild was at night on
Fernandina Island in the Galapagos Islands. It was pitch
black. A low 'haaaaa' split the air. I craned my neck
toward the ocean and the source of the sound. There
was a second 'haaaaa', but it was too dark to see what
made the noise. The next morning I saw a penguin
resting on the shore, preening its feathers — and I fell
in love.

For more than 40 years I've been studying penguins,
but I still feel I have barely scratched the surface of
what they can do, what they need to survive and
what is happening to their populations. Since 1982,
I have studied Magellanic penguins at Punta Tombo,
on the Patagonian coast of Argentina. At the time my
work began, a Japanese company planned to harvest
Magellanic penguins for high-fashion golf gloves, protein
and oil. The New York-based Wildlife Conservation
Society, the Tourism Office of the Province of Chubut,
Argentina, the local landowner and I teamed up to
gather the data needed to make sure no operation
would destroy the penguins that were already a strong
tourism attraction.

Since then, penguins have proven far more valuable
than their products. In the season 2007–
2008 about 120 000 people came to
see the penguins at Punta Tombo. The
challenge now is for both people and
penguins to benefit from this relationship.
By examining a variety of tourism styles, we
found a sustainable model in operations that:
actively manage visitors, provide educational
experiences, generate funds to support
conservation and create local conservation jobs.

Punta Tombo represents the largest
Magellanic penguin breeding colony in the
world, its presence on the South American
continent made possible because sheep farming
has displaced large land predators that would
naturally have occurred in this region. Our
research showed that the deepest dive logged
was 91 m (298.5 ft), but most dives were 30 m
(98 ft) or less. Their lifespan can surpass 30 years,

and we had one pair that bred together faithfully for
16 seasons.

The typical pattern of a penguin's foraging trip,
tracked by satellite, involves meandering when the
penguin leaves the colony, foraging when it is farthest
from the colony, and making a hasty return to the colony
— one covered 173 km (107.5 miles) in a single day.
Returning penguins maintain the same speed of about
7 km/hr (4.3 mph) day and night.

Magellanic penguins favour nesting in burrows, but
these sometimes collapse in heavy rainstorms. I once
dug out a penguin buried alive for nearly two days. He
was very stiff when I found him, but several minutes
after I carried him back to the sea, he stretched and off
he swam.

When food is abundant, Magellanic penguins can
rear two young to fledging in as little as 60 days, but
when food is in short supply this can take twice as long,
and often chicks starve. At Punta Tombo, an extremely
good breeding season may produce roughly one
fledgling from each active nest, but in most years the
average is a little less than one chick per two nests. By
contrast, in the Falkland Islands where food is generally
more abundant, we found reproductive success was
higher, with many nests fledging two chicks.

Our banding records have demonstrated that
Magellanic penguins undertake the longest flightless
migration of any bird — including other penguins —
travelling several thousands of kilometres each year.
Departing their breeding colonies in the Straits of
Magellan, the Falkland Islands and all along the Atlantic
coast of Patagonia, they move northward along well-
defined pathways following their prey, which includes
primarily anchovy, but also squid and hake. Penguins
we have banded at Punta Tombo have travelled north
beyond Rio de Janiero, Brazil, in winter. One penguin
banded as a one-year-old juvenile was found by my
Chilean student as an adult on an island in the Straits
of Magellan.

Magellanic penguins are social and recognise each
other by their calls. When an adult returns to an empty
nest expecting to find a chick, I've seen it bray and the
chick come running to be fed. When we recorded the
parents' calls and played them back in front of the nest,
the chicks came running from their nests and begged
at the loudspeaker. The chicks showed no response
and stayed sleeping in their nests whenever we played

the calls of another pair. Adults likewise recognise their mate and their chicks by their voices.

Threats to Magellanic penguins include: climate change, fishing by-catch, competition with fishers, petroleum pollution, toxic algae blooms and poorly managed tourism. During our 30 years of study, we have tried to cast light on the species' conservation problems. In the 1980s, in some years up to 80% of the penguins my students and I found washed-up dead had petroleum on their feathers, with an estimated annual death rate of some 42 000 individuals. After a decade of accumulated data, government hearings and considerable public outcry, in 1997 the tanker lanes were rerouted 40 km (25 miles) farther offshore. Beach surveys indicate that since that decision was implemented less than 5% of the dead penguins encountered petroleum. While oil pollution is no longer as grave a problem in Chubut, it remains serious from northern Argentina to Brazil.

Fisheries also have their impact, although this is hard to quantify as it varies by region. For example, the penguins' diet composition overlaps with commercial fisheries in areas such as northern Patagonia and the Falkland Islands, whereas hundreds of penguins are known to die yearly in gillnets and trawls in Brazilian fisheries.

We have also been documenting the impact of climate variation using satellite tags on breeding birds. Our penguins today travel about 40 km (25 miles) farther to find food than they did a decade ago. When penguins have to swim farther, their mates on the nest fast and lose weight, so both parents pay the increased energy cost, making it harder for the pair to successfully raise chicks. Concurrently, deadly toxic algae blooms appear to be increasing. In 2000, we lost 12 breeding Magellanics with satellite tags to a toxin. One 18-year-old male was seen just before his mate came back and he went to sea. He was skinny but in good health. His last position was about 150 km (93 miles) from shore, the location of the toxic bloom. We estimated maybe 10 000 penguins died in this one event.

Life is not likely to become easier for Magellanic penguins. Can we humans manage ourselves in such a way that penguins survive and thrive? Their fate rests with all of us. There are many organisations worldwide, such as the Global Penguin Society and others, whose work to save penguins well deserves our support. 🐧

ABOVE Hard glacial soil deposits are honeycombed with nesting burrows, Cabo Dos Bahias, Argentina.
LEFT New arrivals swim up a stream at Seno Otway, Chile.
FAR LEFT Well-fed chicks at their nest entrance.

Southern Rockhopper: Species in Decline

David Thompson

Dr David Thompson is a seabird biologist with the National Institute of Water and Atmospheric Research in Wellington, New Zealand, where he has worked on a range of subantarctic species and islands since 1998. He developed an interest in the application of stable-isotope analyses to seabird studies in the early 1990s while at Glasgow University, and continues with seabird research with a current focus on Campbell Island.

National Institute of Water and Atmospheric Research Ltd.
301 Evans Bay Parade
Hataitai, Wellington,
New Zealand
d.thompson@niwa.co.nz

BELOW With numbers reduced exponentially, colonies are much more susceptible to predation. RIGHT A fledgling chick begs frantically for more food than its parents are able to deliver, Campbell Island.

Synopsis: Using refined technology, a study of stable-isotope levels in penguin feathers produces revealing information about reduced ocean productivity affecting the species.

During cold, grey rainy afternoons years ago in Glasgow, Scotland, talk among the graduate students would often turn to ways we could engineer visits to the world's great seabird colonies of the Southern Ocean. Clearly, Scotland is home to many fine seabird aggregations, but for impressionable young researchers at the start of their careers, the idea of visiting one of the vast penguin colonies of the distant south was altogether more fantastic. That only a few short years later I found myself walking across Campbell Island towards the appropriately-named Penguin Bay was something of a dream come true. Except for one major difference: my destination that day was no longer home to the huge numbers of Southern rockhopper penguins it had once been.

Campbell Island is the southernmost of New Zealand's subantarctic islands, and one of three in the region where Southern rockhopper penguins breed — the other two being Auckland Island to the north-west and Antipodes Island to the north-east. But what distinguishes Campbell from its subantarctic neighbours when it comes to this species, is evidence of one of the most dramatic and large-scale declines in breeding numbers of any penguin and indeed of any marine predator: between the 1940s and 1980s, numbers fell from over 1.6 million to just over 100 000 individuals, a drop of 94%.

The view I beheld from the top of the cliff over-looking Penguin Bay was not of a continuous carpet of breeding rockhoppers, but merely fragmented and isolated sub-colonies separated by zones altogether devoid of birds. It was easy to see where the huge colony had formerly spread, with recovering vegetation now beginning to cover the rocks and boulders. I was also aware that, interestingly, population declines of a similar magnitude had occurred at rockhopper penguin breeding sites in the southern

Atlantic Ocean and, to a lesser degree, in the southern Indian Ocean. There was no doubt that something very profound had occurred here over the course of the preceding few decades. The obvious question was: What?

Phil Moors first reported the decline of Southern rockhopper penguins at Campbell Island in 1986. He and Duncan Cunningham later suggested that changes in sea temperature could be involved, and noted the correlation between rising water temperatures and declining penguin numbers. They found that average sea temperatures during the summer months at Campbell began to rise in the late 1940s and 1950s, fell towards the mid-1960s and rose again during the 1970s. Interestingly, the cooling period in the 1960s coincided with a slight recovery in penguins. These authors speculated that shifting sea temperatures affected the abundance, distribution and availability of penguin prey. Importantly, they also concluded that there was little or no evidence to suggest terrestrial factors such as disturbance by humans or sheep, predation by cats and rats (the three feral species were still present on the island), or that the effects of avian diseases were contributing to these declines. With not a single record of incidental mortality through interaction with commercial fishing activity in New Zealand waters, today we can also discount this as another plausible cause.

To investigate the situation further and explore the findings of Cunningham and Moors, I contacted Geoff Hilton, a fellow graduate student at Glasgow University

with whom I'd shared those early southbound aspirations. He was then working on seabirds in the South Atlantic, and together we developed a plan in 2003 to collect and analyse feather samples — both fresh and historic — from as many rockhopper penguins as possible. This effort included trawling through museum collections around the world to obtain the broadest array of samples from stored specimens, as well as from live birds at far-flung localities. Fortunately, field colleagues working with rockhoppers elsewhere were happy to help, as were the curators of study skin collections. Thus we were able to amass feathers from most breeding sites throughout the species range, and historical samples dating back to the 1860s.

The next step was to determine the ratios of stable isotopes of carbon and nitrogen locked within those feathers. Stable isotopes are variants of elements that differ only in the number of neutrons in the nucleus: for example, carbon-13 has one additional neutron compared to the much more abundant carbon-12. Penguins grow a new set of feathers over a relatively short time frame at the end of each breeding season, following a period at sea when they feed extensively in order to see them through the moulting process ashore. The stable isotope ratios we measured reflected this intensive feeding stint, and contained information about the penguins' diet, including the level of ocean productivity in their foraging habitat. This is because stable-isotope ratios in consumer proteins (in this case, feathers) reflect those in their prey in a predictable manner, and because the carbon isotope ratio of phytoplankton at the base of the food chain depends in part on the growth rate of phytoplankton cells — the more growth or productivity, the higher the carbon isotope signature. This information is transferred up the food chain to the penguins. By analysing a time series of feathers spanning nearly 150 years, we investigated whether rockhopper penguin diet had changed over time, and whether there had been any shifts in ocean productivity that we could link to penguin population trends.

Overall, the results supported the hypothesis of declining ocean productivity through the last century; in short, the oceans were more productive in the past and could therefore support more penguins. In the New Zealand region, this result was particularly strong for Southern rockhopper penguins at Antipodes Island. Although long-term population data here are not as comprehensive as for Campbell Island, there's little doubt over their significant decline in recent decades. Unfortunately, the small number of historical feather samples available from Campbell Island meant that results for this breeding site were inconclusive. Nevertheless, given that Antipodes and Campbell share the same subantarctic water mass and, on an oceanic scale, are in relatively close proximity (about 740 km, or 460 miles), it is reasonable to conclude that both rockhopper populations experience similar ocean conditions.

In conclusion, while changes in ocean productivity appear directly tied to the Southern rockhopper

penguin's decline, it remains plausible that long-term cycling of the Southern Ocean oceanographic climate could still bring about a turn-around in this trend. We have barely begun to tease apart the extremely complex linkages between climate, oceanography — including drivers of water column mixing, like wind strength — and marine productivity. Furthermore, the distinct likelihood that these systems have been perturbed by human-induced global climate change makes the task of identifying natural, long-term fluctuations in Southern Ocean systems even more difficult.

In 2010, we initiated long-term research at Campbell Island that aims to shed new light on the Southern rockhopper penguin's plight, since anecdotal evidence suggests that their decline, quantified up to the 1980s, has continued to the present. Our research will further document their status, including rates of breeding success, adult and juvenile survival and where birds travel to forage, both during the summer breeding months and during the winter when they remain at sea. It is clear that the New Zealand populations of this enigmatic species reflect the fluctuating conditions of its Southern Ocean home. The challenge is to identify and understand the key factors and processes that drive Southern rockhopper penguin population dynamics, and to minimise cumulative anthropogenic impacts. 🐧

ABOVE Coordinated groups riding the surf to shore, Campbell Island.
OPPOSITE TOP Small colonies hide beneath vegetation along the eastern coast of Auckland Island.

BELOW The colony in aptly named Penguin Bay once covered the entire slope, but is now reduced to just a few patches. In just 40 years, between the 1940s and 1980s, the Campbell Island population fell by 94%, from an estimated 1.6 million to just over 100 000 individuals.

Northern Rockhoppers: Tragedy on Tristan

Conrad Glass, MBE

Conrad Glass lives on Tristan da Cunha, with his wife Sharon and son Leon, where he serves as Police Inspector. Born the year the volcano erupted in 1961, he and his family was forced to evacuate, but like many others, they returned two years later. He was elected Chief Islander 2007–2010 by the Island Council and in 2010 was awarded an MBE (Member of the British Empire) for his services to the community. His book, Rockhopper Copper, *recounts the life and times of the people of Tristan.*

Edinburgh of the Seven Seas
Tristan da Cunha Island
South Atlantic
conrad.glass@gmail.com

BELOW On the rocks: the MV *Oliva* aground at Nightingale Island, attended by the MFV *Edinburgh*.

Synopsis: A blow-by-blow report on the oil spill disaster following the grounding of the cargo ship *Oliva* at Nightingale Island: a detailed exposé by the Police Inspector of Tristan da Cunha.

Policemen are often parodied for their love of quoting from their notebook. Here on Tristan da Cunha Island I suppose it was inevitable that once my book *Rockhopper Copper* was published I would become known by the same title. After all, the book does quote at length my notebook entries patrolling the remotest inhabited island in the world — including policing its population of sometimes cantankerous Northern rockhopper penguins.

This was never more so than on Wednesday 16 March 2011: certainly a date for a dramatic entry in the notebook. That day we woke to a grey overcast sky, with wind north-west about 30 knots, and a stormy sea starting to build into breaking waves.

At 08:15 Andy Repetto, from Tristan Radio, informs me a cargo ship has run aground at Spinner's Point on the west side of Nightingale Island. Four hours earlier, the MFV *Edinburgh*, the regular Tristan fishing ship, had responded to a distress call from the MV *Oliva*. Captain Clarence October, from the *Edinburgh*, reported the ship grounded and listing to starboard, the bow about 50 m (164 ft) from the shore.

Nightingale is a small island 40 km (25 miles) south-west of Tristan, covered in long tussock grass, the

breeding habitat for tens of thousands of rockhopper penguins, shearwaters, Yellow-nosed albatrosses and fur seals. About 14.5 km (9 miles) west of Nightingale lies another small island, aptly named Inaccessible for its precipitous contours. Between them, these three islands are home to 65% of the world's Northern rockhopper penguins, an endangered species. The stricken ship was a 75 300-tonne bulk carrier fully laden with soya beans, the MV *Oliva*, en route from Santos in Brazil to Singapore.

From the moment of that phone call, the next few weeks would turn into a cavalcade of events enough to fill several notebooks. By the time the saga finally settled, not only would it involve considerable time and effort by the *Edinburgh* and her crew, but also the help of the passing cruise ship *Prince Albert II* assisting with the rescue of the *Oliva*'s crew, plus two tugboats, the *Smit Amanda* and *Singapore*, making their way at different times from Cape Town with salvage divers, a naval architect, an environmental advisor and specialised gear and experts in the rescue of oiled seabirds. A Russian ship, the *Ivan Pappanin*, would follow with a helicopter, generator plant and equipment to clear the heavy fuel from the shoreline.

Meanwhile, because of grave concern about the risk of ship rats invading pristine Nightingale Island, one of the first decisions by our heads of Conservation and Fisheries Departments, Trevor and James Glass, was to place a ring of rodent traps and poison around Spinner's Point when the first break in the weather allowed. An even more imminent threat would be the inevitable spillage of heavy bunker fuel, precisely at this critical time of year when the majority of penguins were completing their annual moult, ready to take to sea emaciated after three weeks of fasting ashore.

Barely 24 hours later the *Edinburgh* reported the first signs of oil spill. The next day, the scene at Nightingale was awful, with oil stretching 13 km (8 miles) offshore and nearly encircling the island. The slick ranged from a thin film and small balls to larger clumps of tar, with the smell of diesel everywhere. With advice from Trevor Glass and Katrine Herian, project officer for the Royal Society for the Protection of Birds, island Administrator Sean Burns devised a plan to put conservation teams on Nightingale and Inaccessible in order to take oiled penguins away to Tristan to be treated. This was a monumental task: rounding up skittery penguins

©WWW.TRISTANDC.COM

into temporary pens, feeding and placing them into cardboard boxes for transport by the *Edinburgh* on days when the weather permitted. On Wednesday 23 March, the first 19 penguins arrived, the eventual total reaching 3718. The last ones landed on 10 April.

Within five days of starting the rescue we had 1614 penguins to care for. The most urgent task was to rehydrate the weakest ones with electrolytes administered through a tube with a large syringe, to stabilise them before the cleaning could begin. Each penguin would also need about 200 g (0.4 lb) of raw fish daily, which meant securing over 300 kg (660 lb) of fish every day.

The task needed more people to help than the conservation team. Trevor asked for volunteers and the community responded en masse: pensioners, clerks, mechanics, electricians, fishermen, labourers, teachers, nurses, the veterinary officer, heads of departments — all helped. They caught, washed and fed the penguins, erected pens, installed washing facilities and electrical appliances to keep the penguins warm, went out fishing, then cut up the catch into cubes for them. In fact so many volunteered that a rota system was devised so normal fishing and government work could continue, with government staff working alternate days and after hours. The Public Works Department transport building was turned into a rehab clinic, while the police station became the bulk store for the project. When more space was needed, even the public swimming pool was transformed into a penguin paddle pond.

Teams went out in all weathers to catch fish for the penguins, with a crew of five men per each small fishing barge. On days too rough for the boats to set out, the men would use fishing rods from the harbour quay or beach.

Cleaning the penguins involved rubbing vegetable oil into their feathers to loosen the heavy petroleum, and also carefully cleaning oil from their eyes. Then washing could begin. Working in teams of two, the islanders covered the penguins with a mild liquid soap, gently but firmly scrubbing them with nail and tooth brushes, then squirting them with a fine jet of water.

Once cleaned, each penguin was wrapped in a fresh towel and dried, then placed into a clean warm pen to be fed. The next day they would be taken to another pen with small seawater pools on the beach where they were temporarily tagged by putting different coloured

rubber bands on their flippers. There they were fed until they weighed at least 2 kg (6.6 lb) before being released.

On 3 April the first group of 24 healthy penguins was set free. Two days later the tug *Singapore* arrived with more equipment and the SANCCOB team (Southern African Foundation for the Conservation of Costal Birds). Experienced in rescuing oiled penguins, they set up more facilities to help the islanders.

On 21 June the last lot of 180 penguins were finally let go, making a total of 381. For a few weeks some penguins hung around the harbour and would hop ashore looking to be fed. The sad reality is that, in spite of everyone's best efforts, 3337 penguins died at the rehab clinic out of the 3718 brought in for treatment, an 11% survival rate. Over 1000 un-oiled penguins had also been corralled on Inaccessible and released once the oil had dissipated from the shoreline but unknown numbers surely perished around out-of-reach areas of the island.

Even so, one year on from the disaster the prospects for the penguin recovery are looking cautiously optimistic. Winter storms dissipated the oil and a recent survey of rockhoppers returning to nest showed healthy numbers. But the consequences to the marine ecosystem remain complex and insidious, so it may take a long time before we are able to tally this tragedy's full effect. The lobster fishery is the mainstay of the Tristan economy and the impact on the fishing grounds remains a real concern. The Tristan Government, the fishing concessionaire and the *Oliva*'s owners/insurers are still discussing the environmental and economic implications and the best way to manage them going forward.

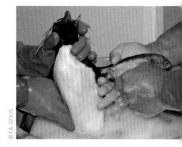

The sad saga of the *Oliva* grounding and ensuing penguin rescue effort, photographed by Tristan Islanders and others directly involved.

PHOTOS COURTESY WWW.TRISTANDC.COM

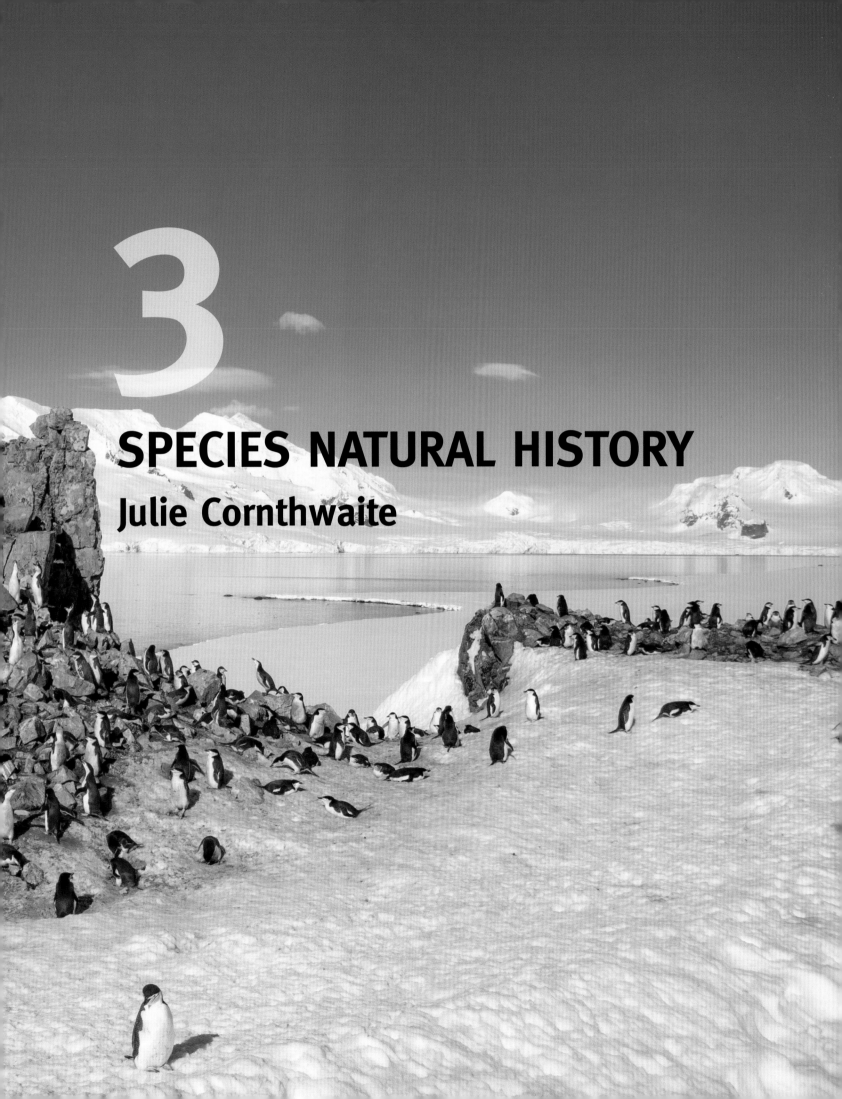

3

SPECIES NATURAL HISTORY
Julie Cornthwaite

Penguin Who's Who: 18 Species Worldwide

KING PENGUIN

EMPEROR PENGUIN

NORTHERN ROCKHOPPER PENGUIN

CHINSTRAP PENGUIN

GENTOO PENGUIN

ADÉLIE PENGUIN

MAGELLANIC PENGUIN

HUMBOLDT PENGUIN

GALAPAGOS PENGUIN

SOUTHERN ROCKHOPPER PENGUIN

ROYAL PENGUIN

MACARONI PENGUIN

ERECT-CRESTED PENGUIN

FIORDLAND PENGUIN

SNARES PENGUIN

AFRICAN PENGUIN

YELLOW-EYED PENGUIN

LITTLE BLUE PENGUIN

Fascinating Penguin Facts

Physical characteristics

Fastest swimmer (Gentoo)
The Gentoo penguin holds the swimming speed record, at 36 kph (22⅓ mph). When travelling fast, all penguins leap clear of the water to breathe without losing momentum, a technique called porpoising.

Him and her (King)
With only a few very minor exceptions, penguins show no gender differences in markings and coloration, but invariably males (here walking on right) are slightly larger and heavier than females.

Non-slip grasp (Gentoo)
All penguins have soft, backwards-pointing barbs lining the inside of their bills, tongues and upper throats, enabling them to grab and swallow slippery prey underwater, without the help of gravity.

Wings as flippers (King)
Penguin wings share the skeletal structure of other birds, except their bones are flattened and rigidly joined, shaping them into smooth, stiff paddles best suited for rapid propulsion in a dense medium like water.

Dual purpose feet (Gentoo)
Robust and versatile, penguin feet have protective scales and gripping claws for secure purchase when walking or clambering on wet and slippery surfaces, from ice to mud and steep rocks. When swimming, webbed feet serve as rudders, with additional flanges on the sides increasing their area.

Feathered beaks (Adélie)
Penguin beaks are powerful and streamlined, slightly articulated for dextrous food handling. In species most adapted to cold climates, such as Adélies, feathers grow partway down the mandibles for insulation.

Underwater flight (Adélie)
Unlike many other diving birds, e.g. ducks and cormorants, penguins swim with their wings while steering with their feet. Rotating shoulder sockets allow enough twist to generate thrust with both up and down wing strokes, a trait shared only with hummingbirds.

Long legs (Gentoo)
Far from their long-standing reputation as clumsy and short-legged, in reality penguins possess long and mobile leg bones hidden within their streamlined contour, while their waddling walk is an energy-saving technique based on the pendulum effect. Some species commute several kilometres to their nests.

Binocular vision (Little blue)
Although penguins often stare one-eyed at objects of interest, studies have shown their forward field of vision is at least as wide as that of owls, important for catching elusive underwater prey.

Adaptable vision (Rockhopper)

Once believed myopic (near-sighted) on land, penguins can actually see remarkably well in both air and water, thanks to very strong focusing muscles and pupils that adjust to very dim light at depth or reduce to pinpricks in brilliant sunlight.

Catastrophic moult (Snares)

Unlike other birds, penguins cannot afford to replace just a few feathers at a time, which would compromise waterproofing. Instead, once a year they must come ashore to replace all their feathers at once, usually just before winter, in what is termed a 'catastrophic moult'. The process takes two to four weeks.

Colour variants (Chinstrap)

Odd-coloured individuals of all species can occasionally be seen, some all black (melanistic), all white (leucistic, not necessarily albino) — as in these Chinstraps — plus blond (isabelline) or mixtures thereof. They can breed successfully with normally pigmented mates.

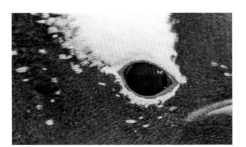

Fading colours (King)

Before becoming land-bound during the annual moult, penguins must fatten-up to obese proportions in preparation for a long fast, their plumage faded and barely recognisable from sun and saltwater wear.

Eye protection (Gentoo)

Like other birds, penguins have nictitating membranes that slide over their eyeballs instead of blinking to lubricate them, especially important underwater.

Super coat (Emperor)

A penguin's plumage is the densest of any bird, with about 15 feathers per square centimetre all over the body, except feet and beak. The silky outer surface helps reduce friction.

Underwater goggles (Galapagos)

Clear nictitating membranes serving as see-through underwater protection, and unusually flat corneas to minimise refraction, give penguins unsurpassed underwater vision.

Unique feathers (King)

With wide, stiff shafts, outer feathers overlap tightly like fish scales, protecting a plush layer of shorter, downy aftershafts in which air is trapped for insulation. When moulting, new feathers push out the old, revealing their true shape as they gradually drop off.

Woolly penguin (King)

With their big shaggy coats giving them an oversize look, large King penguin chicks in their thick winter down were once thought to be an entirely different species of adult penguin, dubbed the 'Woolly penguin'.

Lifestyle

Feather care (Galapagos)
Meticulously looking after their feathers, even at sea, is paramount in order to maintain waterproofing and insulation, including preening and fluffing up the plumage to replenish the air layer trapped within.

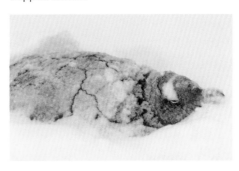

Cool penguin (Gentoo)
Insulation designed to keep penguins equally warm in air and water involves tightly overlapping guard feathers and a downy undercoat, plus a thick fat layer under the skin. Ice on the outside implies there's a cosy penguin within.

Waterproofing agent (Gentoo)
A prominent gland at the base of the tail, called the uropygial or preen gland, secretes a fine oil which penguins carefully pick up with their bills from a special tuft of paintbrush-like feathers, then rub onto their plumage for waterproofing. This can be applied even while at sea.

Hot penguin (Chinstrap)
When exerting ashore in hot weather, such as walking to the colony, penguins are in serious danger of overheating. Bright pink underwings and feet indicate blood flushing to the extremities to cool off.

Air lube (King)
Thin streams of bubbles are shed from penguin feathers during bursts of speed, helping to reduce drag through the water. Eyes adjust to underwater vision by flattening the cornea and enlarging the pupil.

Behaviour

Violent fights (Adélie)
When fighting over mates or territories, penguins bite, kick and pound one another with volleys of hard flipper blows, until one capitulates and runs away. The worst battles happen when a female Adélie returns from her winter wanderings to find another already paired with her previous mate.

Eating snow (Adélie)
During sunny summer weather, at the nesting colony southern penguins may suffer from heat stress and dehydration, often eating snow (if available) to cool down.

Drinking seawater (Gentoo)
Like all seabirds, penguins can drink and process seawater by means of large salt-extracting glands in their foreheads serving to remove excess salt from the bloodstream and excreting it in concentrated form.

Nest building (Gentoo)
Unlike other birds (e.g. the cormorant seen here), penguins don't carry nesting material back over long distances, but use what's available nearby, such as pebbles, that are important for drainage of meltwater and mud that could damage the eggs.

Chick feeding (Chinstrap)
To feed their chicks, all penguins carry food stored in their stomachs, which they regurgitate beak to beak. Krill figures high in the diets of many southern species, such as this Chinstrap, but squid and small fish are equally important prey, especially in more temperate species.

Baby carrier (King)
Kings and Emperors are unique in carrying their eggs and chicks on top of their feet, covered by a warm 'blanket' of loose feathered skin. They can even shuffle about without losing their precious cargo.

Crèche guardian (Rockhopper)
In many of the smaller species, such as Rockhoppers, parents leave their half-grown chicks huddling together for safety while they are out getting food for them. Called a 'crèche', such groups are often guarded by one or two non-breeding neighbours.

Ecstatic courtship (Snares)
With some variations between species, all penguins carry out exuberant courtship displays involving many expressive postures and loud calls, of which the most striking are the sky-pointing and 'ecstatic greeting', often performed when partners reunite after returning from the sea.

Threats

Deadly epidemics (Gentoo)
Pathogens, such as avian cholera and avian pox (seen here in a Gentoo chick, Falkland Islands), occasionally cause virulent epidemics, but whether this is spread naturally among seabirds or if human visitation plays a role has not been fully determined.

Predator attacks (King)
Leopard seals, sea lions and fur seals are all wily natural predators that often stalk penguins near their landing spots, deftly shaking them out of their skins to feast on blubber and breast muscles. This King got away but may not survive.

Oil pollution (Little blue)
Since they cannot fly to avoid even the smallest of spills, penguins are particularly vulnerable to oil worldwide, but especially near shipping lanes and oil wells. Petroleum, even in small quantities, kills by destroying feather waterproofing and by its toxicity when ingested while preening.

Penguin Range and Population Status*

GENUS & SPECIES	STATUS **	TOTAL POPULATION ESTIMATES AND TRENDS	RANGE AND BREEDING SITES	MAIN THREATS
Aptenodytes		From Ancient Greek 'a' without, 'pteno' able to fly or winged, and 'dytes' diver. Largest and perhaps most ancient of living penguins, sharing features with many extinct giant penguins. Highly colonial, lay single egg incubated on top of feet, under loose flap of abdominal skin. Pelagic deep divers.		
Emperor *Aptenodytes forsteri*	LC	256 500 breeding pairs (satellite survey 2009, updated 2019); decline in some colonies projected to accelerate.	Circumpolar, feeding from ice edge well into open waters of Southern Ocean; 54 nesting colonies on winter sea-ice dotted near continental shores.	Breeding failure due to rising sea temperatures and increasing melt of Antarctic ice causing habitat loss and reduced food availability.
King *Aptenodytes patagonicus*	LC	1.1 million annual breeding pairs; most colonies increasing except at northern limit of range.	Widely distributed around southern oceans, nesting on islands near Polar Front with largest colonies: South Georgia, Crozet, Kerguelen, Macquarie.	As oceans warm, increasing commuting distances between nesting and feeding grounds may become too great for chick provisioning.
Pygoscelis		'Pygoscelis' means 'rump-tailed' in Greek. Commonly called the long-tailed, or brush-tailed, penguins for their very long, stiff tail feathers, these are true Antarctic species — largely krill and ice-dependent, except Gentoo feeding on bottom fish and also found around subantarctic islands. All have partially feathered bills, most notable in Adélie. Lay two eggs in nest mound made of pebbles or other debris. Both chicks stand good chance of fledging.		
Adélie *Pygoscelis adeliae*	LC	3.79 million breeding pairs; stable/increasing, but modelling projections suggest mid-century decline.	Circumpolar in broken pack-ice, nesting along Antarctic coast and adjacent islands, with largest numbers in Ross Sea region.	Climate change with associated rising sea temperatures and excessive precipitation (severe snowfall or rain) reducing sea-ice habitat and breeding success.
Chinstrap *Pygoscelis antarcticus*	LC	Minimum 8 million mature individuals estimated but local population trends are complex across range, with reliable figures lacking for many areas; growth of past decades appears reversed.	Large circumpolar pelagic range along seasonal ice edge; breeding centred mainly on islands around Antarctic Peninsula and Scotia Arc. Some new colonies being established following decrease in extent of sea ice.	Reduction of sea-ice cover and abundance/distribution of krill (key prey species), due to climate change, believed to drive recent population crash in some colonies; expanding krill exploitation potential for further stresses; some colonies at risk from natural volcanic activity.
Gentoo *Pygoscelis papua*	LC	774 000 mature individuals; assumed stable with some regional fluctuations; general increase/expansion in southern range.	Widest latitudinal range in genus, with colonies throughout southern oceans, from snow-free Falkland Islands well north of Polar Front to ice-clad continental shores of Antarctic Peninsula.	Disease epidemics and biotoxins (avian pox and red tide) may cause significant local mortality events; colonies easily disturbed, impacting breeding success; actively feeds on trawling fleet discards so susceptible to bycatch from fisheries activities.
Spheniscus		From Greek, 'Spheniscus' meaning 'wedge'. Temperate and tropical species, centred around South America and South Africa, feeding mainly on small schooling fish. All have varying configurations of classic black-and-white banded pattern around neck and chest. Shows least cold adaptation, with bare skin around face and eyes. Cavity nesters, excavating ample burrows depending on substrate available. Raise two chicks when food supply permits.		
African *Spheniscus demersus*	EN	Around 17 700 pairs (2019); rapid decline on-going at most South African colonies, declining more slowly in Namibia.	South African and Namibian coastal feeding habitat concurrent with sardine stocks; breeds at 22 colonies (18 in South Africa and 4 in Namibia).	High mortality from major oil spills; intense competition from fisheries, plus offshore shift of sardine and anchovy banks in response to ocean warming. Also habitat loss for coastal housing and port developments.
Magellanic *Spheniscus magellanicus*	LC	1.1 to 1.7 million pairs; population trends vary across range but considered relatively stable or slowly declining overall.	Mainly Patagonia (with huge dense nesting colonies) and Falkland Islands (widespread), with northward winter migration to southern Brazil; smaller numbers nest along Pacific coast to central Chile.	Climate change bringing either drought or increased rainfall affecting habitat suitabillity; oil pollution from Patagonian offshore wells and tanker traffic, and further development planned around Falkland Islands; competition from anchovy fishery, plus localised bycatch and use as bait in crab pots.
Humboldt *Spheniscus humboldti*	VU	Massive declines in 19th and 20th centuries; population estimated at 23 800 mature individuals; probable continuing decline in some regions; further assessments needed with consistent data collection methods.	Fairly sedentary, endemic to coastal upwelling of Humboldt Current, central Chile to northern Peru. Nests on small desert islands (and some protected headlands) along Sechura-Atacama Desert coast.	Historically, suffered tenfold decline from guano mining removing burrowing substrate, competition from fishing industries, plus starvation during severe ENSO events. Currently artisanal fisheries bycatch (entanglement and explosives) and targeted use (bait and food). Oil pollution from shipping an increasing threat.
Galapagos *Spheniscus mendiculus*	EN	1042 individuals counted (last census, 2009); long-term decline with sharp mortality during prolonged El Niño events.	Galapagos Islands, concentrated around cold upwelling areas; nesting mainly on Fernandina and western Isabela Islands.	Climate change, with increased severity of El Niño events causing prey dispersal, breeding failure and starvation; introduced predators (mainly cats); potential epidemic from alien pathogens; plastic marine pollution.

* information based primarily on Birdlife International Species Factsheets, June 2021

**EN: Endangered VU: Vulnerable NT: Near threatened LC: Least concern

GENUS & SPECIES	STATUS **	TOTAL POPULATION ESTIMATES AND TRENDS	RANGE AND BREEDING SITES	MAIN THREATS
Eudyptes	'Eudyptes' from Greek, meaning 'good diver'. Most diversified genus, with wide oceanic distribution, nesting on islands north of Antarctic Convergence zone. Several species restricted to small range or single island, e.g. Snares and Macquarie. All scale steep, rugged terrain — or nearly sheer cliffs — to access dense, noisy inland colonies, except Fiordland whose colonies are small and scattered under forest cover. Lay two-egg clutches, with second egg much larger than first; smaller chick nearly always lost due to parental neglect or sibling competition.			
Southern rockhopper *Eudyptes chrysocome*	VU	1.27 million pairs, though some counts outdated, with drastic declines in most areas over a 40–70 years span. Estimated trend over last 3 generations is over 30% decline, expected to continue at a similar rate.	Found in all sectors of southern oceans, between 46–54°S; main breeding islands off southern South America (esp. Falklands), South Africa, Australia and New Zealand, also French Crozet and Kerguelen.	Poorly understood threats include decreasing ocean productivity, linked to climate change, responsible for historic crash. Biotoxins from algal bloom caused mass mortality event in Falklands (2002/3). In 2016 another mass mortality (total extent unknown) in South Atlantic mainly due to starvation over the moulting period.
Northern rockhopper *Eudyptes moseleyi*	EN	Approximately 206 850 pairs. Considered stable in northern islands of Atlantic Ocean but continuing to decline on Gough Island (e.g. 98% in 51 years, 1955–2006).	Limited to central South Atlantic and southern Indian Oceans; roughly 85% breed on Gough and Tristan da Cunha with remainder on Amsterdam and St Paul Islands.	Threats poorly understood, with no verified explanation for severe decline at most locations. May include fisheries competition, human disturbance/predation, climate-related fall in ocean productivity, or combinations thereof.
Macaroni *Eudyptes chrysolophus*	VU	6.3 million breeding pairs estimated worldwide; believed to be declining sharply in some areas.	Widely distributed in southern Atlantic and Indian Ocean sectors south of Polar Front, nesting at approx. 55 sites.	Vulnerable to climate-induced changes in food supplies; disease outbreaks recorded in some colonies (Marion Island); competition for prey and breeding habitat from fur seals.
Royal *Eudyptes schlegeli*	NT	850 000 estimate (1984/5) should be updated; believed stable.	Endemic to Australia's subantarctic Macquarie Island and adjacent waters.	Though well protected, limited range highlights vulnerability to potential catastrophic event (e.g. epidemic, oil spill, etc.) and climate-induced food supply shift.
Fiordland *Eudyptes pachyrhynchus*	NT	12 500 to 50 000 mature individuals, although population trend still considered to be decreasing.	Found only along southwest coast of New Zealand, breeding in small secretive colonies in coastal forest and adjacent small islands.	Introduced predators; accidental fisheries impact and some human disturbance. El Niño climate events shown to affect breeding success in some West Coast locations.
Snares *Eudyptes robustus*	VU	25 149 breeding pairs counted in 2013; considered stable.	Nests only on Snares Islands, feeding nearby when breeding; winter dispersal to subtropical Indian Ocean, north of 45°S.	No current threats, but potentially vulnerable to climate change, fisheries and oil spills due to restricted breeding range.
Erect-crested *Eudyptes sclateri*	EN	c. 150 000 mature individuals, presumed decreasing.	Restricted to New Zealand subantarctic Bounty and Antipodes islands; non-breeding dispersal unknown.	With no verified threats, presumed declines attributed to 'marine factors affecting survivorship' (Birdlife International 2020).
Megadyptes	'Megadyptes' from Greek words meaning 'big diver'. Range restricted, single species genus widely separated from all others. Very shy, nesting under dense vegetation in widely dispersed colonies without direct line of sight between nests. Winter breeder, with two chicks often raised. Near-shore, bottom feeder; non migratory.			
Yellow-eyed *Megadyptes antipodes*	EN	Estimated 2684 to 3064 mature individuals; probably declining but difficult to assess, with substantial fluctuations recorded.	Endemic to New Zealand south-eastern coast, Stewart Island and subantarctic Auckland and Campbell islands. Sedentary around breeding sites, except for some northward movements by juveniles.	Numerous threats include: introduced and natural predators (mainly cats, pigs, dogs; rogue sea lions); habitat loss (fire and vegetation clearance); disease (digestive bacteria and blood parasites); human disturbance; fishing net bycatch.
Eudyptula	'Eudyptula' derives from Greek, meaning 'good little diver'. Smallest of family, divided in several subspecies, particularly New Zealand population, with White-flippered penguin of Banks Peninsula sometimes considered separate species. Nocturnal and mildly colonial nester, in burrows, caves or rocky/vegetated habitat. Raises two chicks and may double clutch when food permits. Coastal feeder, mostly small schooling or bottom fish; non migratory.			
Little blue *Eudyptula minor*	LC	Estimated 469 760 breeding pairs, but very difficult to ascertain due to scattered distribution; considered stable overall.	Inshore distribution with small nesting colonies along southern coasts of Australia, Tasmania and New Zealand; sedentary, with no marked seasonal movements.	Predation by introduced mammals; road kills where nighttime traffic passes between shoreline and nest sites; human disturbance; coastal development causing major loss of habitat; climatic variability increasing land temperatures and reducing prey availability.

Emperor Penguin
Aptenodytes forsteri

Alternative or previous names: None known.
First described: G.R. Gray, 1844.
Taxonomic source: Christidis and Boles (1994, 2008); SACC (2006); Sibley and Monroe (1990); Stotz et al (1996); Turbott (1990).
Taxonomic note: No recognised subspecies.
Origin of name: Specific '*forsteri*' after Johann Reinhold Forster, German naturalist with Captain James Cook on his second voyage and one of the first to describe penguins, officially naming five species.
Conservation status: Near threatened (IUCN 2020 — uplisted in 2012).
Status justification: Although many uncertainties exist regarding climate change forecasting, a rapid increase in the rate of population decline is predicted within a couple of generations.

ABOVE Sharp toenails are used as crampons to travel over slippery ice, Prydz Bay, East Antarctica.
ABOVE RIGHT Pair courting, Atka Bay, Weddell Sea.

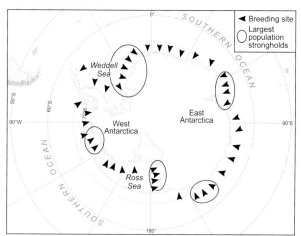

ABOVE Chick moulting into fledgling plumage, beginning to show juvenile markings, Cape Darnley, East Antarctica.
BELOW Using tide cracks in sea-ice to access open water, Emperors can navigate and travel considerable distances beneath the frozen surface, Kloa Point, Edward VIII Gulf.

SIZE AND WEIGHT
Body length 100–130 cm; weight variable depending on gender and time of year, ranging between 22–40 kg.

VOICE
Similar to King, varying in duration and syllables, with two notes emitted simultaneously. Relies on voice recognition with trumpeting call emitted by sky-pointing bird on return to colony to locate partner or chick. In courtship, call is performed by both sexes and is more complicated and rhythmical duet, punctuated by periods of silence. Once pair formed, tend to remain silent until after egg laying. Chicks and small juveniles emit variable pitched, short or prolonged whistles.

POPULATION AND DISTRIBUTION
Circumpolar, between 54° and 78°S. Most ice-adapted of any penguin species, usually breeds on stable fast-ice surrounding Antarctic continent and adjacent islands, with major colonies in Ross Sea region, Weddell Sea and East Antarctica.

Considered dispersive after breeding season, mid/late December to April, but little known about movements. Breeders travel between 150–1000 km per foraging trip. Fledglings tracked by satellite show strong tendency to head north into open water, some travelling

DESCRIPTION
The largest of all penguins and only one to breed during Antarctic winter, enduring harshest conditions of any species. Confusion with slightly smaller and more colourful King penguin possible but differs significantly in reproductive biology, range and habitat. Normally breeding on sea-ice, only three colonies recorded on land — thus many Emperors spend their entire lives without ever touching terra firma, a trait unique among birds.

COLORATION
Adult: Jet black throat and head, adorned with golden ear patches fading to pale lemon down sides of neck to upper breast. Black band down side of body separates white ventral area from steel-grey back, with black tail. Bill slender with slight downward curve, black upper mandible, lower mandible peach-orange from base turning lilac towards black tip. Very dark brown eyes. Black feet.
Immature: Similar to adult but slightly smaller and slimmer. Paler back, head and around eye, off-white or silver-grey throat and less defined, whitish ear patches. Black bill.
Chick: Body covered with silver-grey down, head black with white 'goggle-like' mask around eyes and across throat.

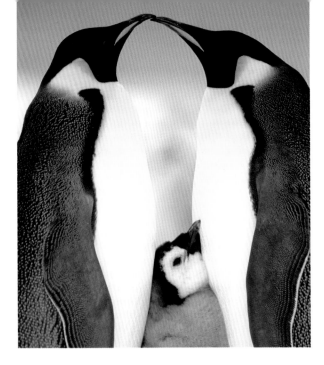

over 1500 km to 54°S (see Wienecke, p. 176).

Recent population reassessment using 2009 satellite imagery, updated in 2019, produced much higher results than previously published figures, showing 54 colonies with an estimated 256 500 breeding pairs.

Vagrant to South Shetland, Falkland, South Sandwich, Kerguelen and Heard islands, Tierra del Fuego and New Zealand.

BREEDING
In March/April, males undertake famed 'march' across winter sea-ice to establish breeding colonies, followed shortly by females. Often travel in long lines, walking or 'tobogganing' on their bellies and covering distances between 50–120 km. Colonies usually sited on level fast-ice, amid grounded icebergs, islands, capes or glacier tongues which help stabilise sea-ice and provide wind shelter. Start breeding at 5–6 years, rarely age 3. Low pair fidelity season to season.

Courtship: Commences with male standing motionless with head on chest, inhaling and emitting call with head lowered, repeated at different locations within colony until selected by female. Bonding pair continues with rhythmic trumpeting duet, joint sky-pointing, holding position for several minutes, moving around colony together with exaggerated swaying walk and bowing. Evolves to mutual displays and alternate trumpeting.

Nests: No nest built. Egg held on top of feet and incubated against bare brood patch, covered with sagging fold of feathered abdominal skin.

Laying: May to early June; highly synchronous within colony. Single, pale greenish-white egg passed to male soon after laying. This hazardous manoeuvre performed quickly to prevent egg freezing.

Incubation: 62–67 days. Males solely responsible while females leave colony to forage. Males survive harsh winter conditions by forming tight huddles, constantly shuffling in a circular pattern moving gently toward the centre to avoid continuous exposure, but may lose 45% body weight during four months fast.

Chick rearing: Eggs hatch mid-July to early August. Females time their return to take over from males at hatching but, if late, males are able to feed hatchlings for up to 10 days on special protein-rich, milk-like stomach

secretion (see Le Maho, p. 160). Chicks brooded 45–50 days on parents' feet in alternating shifts, then left to form crèches. As chick grows, feeding becomes more frequent in tandem with receding pack-ice.

Fledging: 150 days, mid-December to early January, coinciding with break-up of fast-ice.

FOOD
Forages in open sea or through breaks in sea-ice, on small fish, krill and squid. Capable of submerging up to 20 minutes. Prefers shorter dives, usually at depths around 100–120 m but capable of diving beyond 400 m; deepest documented dive to 564 m (see Wienecke, p. 176).

PRINCIPAL THREATS
Predators: At sea, orca and leopard seals. At colony, Southern giant petrels and South polar skuas kill chicks and scavenge corpses.

Fisheries: Fish, krill and squid all subject to some level of commercial harvest in foraging areas. Expansion of krill and silverfish fisheries (both part of their diet), coupled with global warming impacting stocks, could affect long-term species survival.

Climate change: Increase in global temperatures and subsequent reduction in extent, thickness and duration of seasonal sea-ice habitat, with possible changes to ocean circulation, likely to pose major threat to both breeding success and food supply.

Disturbance/Tourism: Potential human disturbance, especially aircraft movements and proximity to science facility operations, may impact some breeding colonies. 🐧

ABOVE LEFT Rarely do both parents meet while feeding their chick, Atka Bay, Weddell Sea.
ABOVE Long distance travel over the sea-ice, Edward VIII Gulf.
LEFT Tobogganing takes little energy; shiny breast feathers gliding with ease over ice and snow, Seymour Island, Weddell Sea.

BELOW Parent with 3-month-old chick, Prydz Bay, East Antarctica.
FAR LEFT About six weeks from fledging, chicks outnumber adults at the Amanda Bay colony, Prydz Bay, East Antarctica.

King Penguin
Aptenodytes patagonicus

Alternative or previous names: 'Oakum Boys' (chicks) so named by old sealers who likened them to the young seamen who became covered in the stripped old rope and tar they used to re-caulk ships in England. Downy chick historically referred to as Woolly penguin.
First described: *Aptenodytes patagonia* J.F. Miller, 1778.
Taxonomic source: Christidis and Boles (1994, 2008), Dowsett and Forbes-Watson (1993), SACC (2005 + updates), Sibley and Monroe (1990,1993), Stotz et al. (1996), Turbott (1990).
Taxonomic note: Two subspecies recognised: *Aptenodytes patagonicus patagonicus* (J.F. Miller, 1778) breeding in South Atlantic; *Aptenodytes patagonicus halli* (Mathews, 1911) in Indian Ocean and southwestern Pacific. Several studies have shown significant size variations in different island populations. Evidence also points to genetic differences between those on Crozet and Kerguelen Islands (C.R. Viot, 1987).
Origin of name: *patagonicus* derived from Patagonia, in turn meaning 'big foot' or giant.
Conservation status: Least concern (IUCN 2020 — since 1988).
Status justification: Extensive range, with large, increasing populations (except at northern limit of range), following historic exploitation.

BELOW AND BELOW RIGHT
The sky-pointing display, with double-noted trumpeting, is used in courtship and for partner recognition, Volunteer Beach, East Falkland.

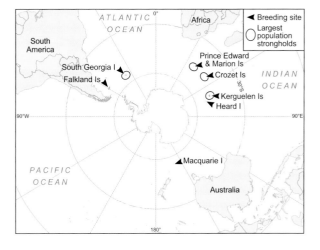

DESCRIPTION
Second largest and most colourful penguin. Confusion with larger and paler Emperor penguin possible, but differs significantly in reproductive biology, range and habitat.

COLORATION
Adult: Silvery grey-blue back, with black tail feathers. Black head adorned with bright orange ear patches and upper breast, fading into white ventral area. Bill long and slender with slight downward curve, black with broad bright orange or pink stripe along lower mandible. Dark brown eyes. Black feet.
Immature: Slightly smaller than adult, with similar plumage but paler orange markings. Greyish-white throat, paler bill stripe. Adult plumage acquired at the beginning of third year.
Chick: Naked and nearly black at hatching, first growth of short, dense grey-brown down is replaced by shaggy, cinnamon-brown, woolly-looking coat retained until fledging. At this stage, mistaken by early explorers for an entirely different species of 'woolly penguins'.

SIZE AND WEIGHT
Some population variations. Body length 85–95 cm; weight variable depending on gender and time of year, ranging between 9.3–17.3 kg.

VOICE
Calls consist of two notes emitted simultaneously, varying in duration and syllables depending on geographic location; female somewhat higher pitched. Contact call a short trumpeting, emitted by birds returning to colony to seek partner or chick. Courtship call a loud, melodious trumpeting, with bill pointing skywards — usually short during pair formation, longer in well-established pairs. Longer calls also used for partner and chick recognition. Grunts used in aggression, especially between incubating neighbours. Chicks and juveniles emit soft warbling whistles.

POPULATION AND DISTRIBUTION
Breeds on most subantarctic islands between 45° and 55°S, with total population approximately 1.1 million annual breeding pairs. Largest colonies on South Georgia Island (c. 450k), Crozet (377k), Kerguelen (377k), Macquarie (150–170k); smaller numbers on Marion/Prince Edward, Heard and Falkland Islands.

Some birds present in all colonies throughout year. When feeding small chicks, parents have been tracked 80–418 km from colony, but appear to migrate considerable distances in winter, when chicks are left to fast for up to 3 months. Odd pairs attempt breeding on South Sandwich Islands and South Shetland Islands.

Vagrant to New Zealand, South Australia, South Africa, Gough Island and Antarctic Peninsula.

BREEDING
Intensely social, forms very large, noisy colonies sited on level ground

or gentle slopes, free of snow and ice, with beach access. Asynchronous breeding cycle takes 14–16 months, so may reproduce successfully only twice in three years, or in some areas, biennially. Although monogamous for the season, pair fidelity between years is low. May start breeding at 3–4 years, but most commence around age 5–8. Chances of winter survival are poor for small chicks, so late starters that fail become early breeders in following season.

Courtship: Courtship, often performed in groups, is stately, starting with male ecstatic displays, standing very erect, sky-pointing with neck extended and emitting trumpeting call. Bonding pair continues with synchronous sky-pointing and bowing, bill-shaking and clapping, and exaggerated strutting, swinging heads from side to side. Advanced courtship evolves to joint sky-pointing, long duets of rhythmical calls with a final, abrupt descending note accompanied by a rapid downwards jerk of the head.

Nest: No nest is built. Egg held on top of feet and incubated against bare brood patch, covered by sagging fold of feathered abdominal skin.

Laying: November to March; variable due to long breeding cycle. Single white, pear-shaped egg has soft, chalky surface which hardens and becomes pale green within days. Female passes egg to male a few hours after laying.

Incubation: Average 52–56 days, shared by both parents in shifts of 12–21 days.

Chick rearing: Brooded for 31–36 days, balancing on parents' feet and sheltered by brood pouch. After initial care phase, chicks form crèches, being fed every 5–7 days, but only sporadically in winter May to September/October, losing weight until feeding resumes in spring.

Fledging: Between 10–13 months, but highly variable depending in part on hatching date. Main departure dates end December to late February, but varies with location.

FOOD
Mostly mid-water pelagic fish, especially small biolumi-nescent lanternfish and some squid at depths of 50 m or less, but at times to 100–300 m.

PRINCIPAL THREATS
Human exploitation caused major declines throughout range in late 19th/early 20th centuries, harvested for oil, plus egg collecting. Populations on Heard, Falkland Islands and one colony on Macquarie Island, were totally exterminated but over time have recolonised.

Predators: At sea, mainly orca, leopard seals, plus adult male Antarctic fur seals and South American sea lions, the latter two sometimes also hunting on land. Ashore, Giant petrels attack juveniles and chicks, skuas take small chicks and eggs if unattended. Feral dog attacks have occurred in the Falkland Islands.

Fisheries: Occasional entanglement in net discards.

Climate change: Vulnerable to environmental changes, with parents forced to travel greater distances to forage due to warming seas around breeding islands, causing stress in chick provisioning. This scenario suspected as one reason for once largest breeding location on Île aux Cochon, Crozet Islands suffering 88% population decline over 35 years (Weimerskirch et al. 2018).

Oil pollution: Potential spills from expanding oil exploration and drilling in Falklands waters. ⬆

Adélie Penguin
Pygoscelis adeliae

Alternative or previous names: None known.
First described: *Catarrhactes adeliae* Hombron and Jacquinot, 1841.
Taxonomic source: Christidis and Boles (1994, 2008); SACC (2006); Sibley and Monroe (1990); Stotz et al (1996); Turbott (1990).
Taxonomic source: No recognised subspecies, though Ross Sea population is genetically distinct.
Origin of name: *'adeliae'* after Terre Adélie, in turn named by French explorer Jules Dumont d'Urville in honour of his wife, Adèle.
Conservation status: Least concern (IUCN 2020 — downlisted 2016).
Status justification: Net change in world population, numbering in the millions, considered as stable or recently increasing throughout most of the breeding range.

ABOVE RIGHT A first-year immature in the open pack-ice; differs from adults by its white throat.
RIGHT Small groups resting on drifting ice floes, Weddell Sea.

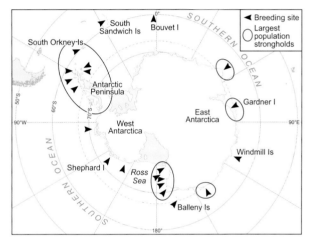

BELOW Crest and sclera (white of eye) displayed in courtship, Gardner Island, Prydz Bay, East Antarctica.
RIGHT Fledgling begging parent for food, King George Island, South Shetlands.

DESCRIPTION
Southernmost of all penguins, breeds to 77°33'S during brief southern summer. Feisty and aggressive when defending territory. Subfossil Adélie penguin remains from Ross Sea region are the oldest known of any extant penguin species, aged 45 000 years.

COLORATION
Adult: Black head, throat, back and tail, with completely white ventral area. Able to raise feathers at back of head to form crest. Dark brown eyes, almost black, with distinctive white eye-ring plus white sclera (white of eye) visible during certain displays. Bill appears short due to black feathering extending over much of both mandibles, visible part black with dull orange-red markings. Pale to dark pink feet with black soles.
Immature: Similar to adult but slightly smaller and slimmer. Dark areas more blue-black, with white throat; white eye-ring absent.
  Chick: Initially silver-grey body with darker grey head, then completely dark sooty-grey.

SIZE AND WEIGHT
  Body length 70–73 cm; weight variable depending on gender and time of year, ranging between 3.8–8.2 kg.

VOICE
Similar to Chinstrap, but more forceful, mostly during territory formation and courtship. Contact call loud and sharp, with slight rise in pitch over short duration. Display call a repetitive staccato similar to braying, exhaling and inhaling while pumping flippers. Growls and grunts during territory defence or fighting. Chicks emit repetitive peeps, or whistle when begging.

POPULATION AND DISTRIBUTION
Circumpolar, along Antarctic coast and adjacent islands, from Cape Royds (77°33'S) in Ross Sea, north to Bouvet Island (54°25.8'S). Global population higher than previously estimated, now at about 3.79 million breeding pairs; with all age cohorts considered, the total population could be as high as 14–16 million (Southwell et al. 2017). Recently surveyed megacolonies found in Danger Islands, N. Weddell Sea (Borowitz et al. 2018) contains 750 000 pairs, more than the rest of the entire Antarctic Peninsula region combined. Range overlaps with Chinstraps and Gentoos in places.

Considered dispersive, absent from colonies May to August, migrating northwards to outer edges of pack-ice to forage and moult after breeding. Satellite tracking found Ross Island Adélies make longest migration known for this species, about 17 600 km round-trip (Ballard et al, 2010). Vagrant to Australia, New Zealand, Argentina, Falkland Islands and subantarctic islands in Indian and Pacific oceans.

BREEDING

Gregarious and territorial, breeds in dense colonies established on accessible ice-free coastal slopes and headlands, beaches and rocky islands, numbering from a few hundred to over 300 000 (Cape Adare). Earliest age for breeding between 3 and 5 years. Season is short, males returning to colonies between September and October, shortly followed by females. Strong nest site fidelity, vigorously defended. Monogamous, often with same partner in consecutive years.

Courtship: Commences with ecstatic display, usually by single male at nest site, stretching upwards, rolling eyes down and back, raising crest and sky-pointing, emitting repeated and initially soft but increasingly loud pulsating call with rhythmical flipper movement. May follow this with 'bill to axilla' display, turning head to point bill under flipper. Bonding pairs perform mutual ecstatic display, with loud calling duets or lower pitched calls, waving heads and bowing.

Nest: Shallow scrape in frozen earth lined and rimmed with stones (important for meltwater drainage). Competition for pebbles is high, and stealing from other nests common.

Laying: Highly synchronous. Normally two egg clutch in October/November, 2–4 days apart. The first A-egg is larger than the second B-egg. Single egg may be laid by young breeders.

Incubation: 32–37 days. Full incubation commences after second egg is laid, either crouching forward with eggs on feet, or by lying down in nest. Shared by both parents, usually male first for about 2 weeks, while female forages, then similar foraging period for male. Shorter shifts follow until eggs hatch. Breeding failure often caused by inability to coordinate changeover, with late return (possibly caused by food shortages or long journeys to open water) forcing incubating bird to abandon in order to survive. No replacement clutch.

Chick rearing: Eggs hatch about one day apart, both chicks may be raised. Shared equally by both parents. Chicks form crèche after about 20–30 days, when both parents leave to forage, returning every 1–3 days to provision chick. Feeding of larger chicks often involves food chases before feeding session.

When food scarce only stronger chick in clutch survives.
Fledging: 50–56 days.

FOOD

Mainly krill, fish and some squid, depending on location and seasonal availability, usually at depths of 20 m or less but is capable of diving up to about 175 m.

PRINCIPAL THREATS

Predators: At sea mainly leopard seals concentrating attacks at edge of pack-ice where penguins enter and leave the water. Occasionally taken by orca. On land, giant petrels and skuas take unguarded chicks and eggs.

Fisheries: Krill and squid subjected to some level of commercial harvest in foraging areas may result in resource competition.

Climate change: Likely negatively impacted by shrinking sea-ice cover; in some areas, excessive precipitation (severe snowfall or rain) caused by warmer temperatures disrupts breeding success (See Ainley, p. 166).

Pollution: Large colonies at increased risk of potential oil spills from ships servicing both nearby research stations and the tourism industry.

Disturbance/Tourism: Increased tourism may facilitate skua predation; logistical operations using fixed-wing aircraft and/or helicopters may affect some colonies, especially those located close to large scientific bases. ↟

TOP RIGHT Colony surrounded by sea-ice during incubation, Gardner Island, Prydz Bay, East Antarctica.
TOP LEFT A group of Adélies rides a fast-melting ice floe near Possession Island, Ross Sea.
ABOVE LEFT Pebble-lined nest in the snow, Paulet Island, Weddell Sea.
ABOVE Incubating adult trapped by heavy snowfall, Devil Island, Weddell Sea.
BELOW Stone gathering is important for nest lining, Gardner Island, East Antarctica.

Chinstrap Penguin
Pygoscelis antarcticus

Alternative or previous names: Ringed, Bearded or Stonecracker penguin.
First described: *Aptenodytes antarctica* J.R. Forster, 1781.
Taxonomic source: Christidis and Boles (1994, 2008); SACC (2006); Sibley and Monroe (1990); Stotz et al (1996); Turbott (1990).
Taxonomic note: No subspecies recognised. Gender agreement of species name (*antarcticus* versus *antarctica*) follows David and Gosselin (2002b).
Origin of name: *'antarcticus'* after the region it inhabits; common name derived from facial markings.
Conservation status: Least concern (IUCN 2020 — since 1988).
Status justification: Very large range and population size, well outside 'Vulnerable' status thresholds. Appears to be increasing at southern extreme of range; but declining at most Antarctic Peninsula sites; stable at South Sandwich Islands.

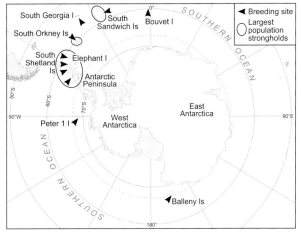

TOP LEFT AND RIGHT Males sky-pointing, pumping their wings and braying, to advertise for a mate.
ABOVE Eating snow to cool down during hot weather.

BELOW Carrying nest-building pebble.
RIGHT Resting on wave-worn iceberg beyond the pack-ice edge, South Orkney Islands.

DESCRIPTION
Extremely agile, bold and feisty penguin. Species most often seen riding isolated, wave-worn icebergs in open ocean. Confusion with other species impossible due to unique markings.

COLORATION
Adult: Blue-black crown, forehead, nape, back and tail. White from above and around eyes and bill, to throat and ventral area. Easily identified due to fine black line passing under chin from ear to ear. Amber eyes rimmed with black. Black bill. Pink feet with black soles.
Immature: Closely resembles adult but slightly smaller and slimmer. Dark areas paler, with dark spotting on face, especially around eyes; black eyes.
Chick: First down pale silver grey, then darker grey head and back with slightly paler ventral area.

SIZE AND WEIGHT
Body length 68–76 cm. Weight variable depending on gender and time of year, ranging between 3.2–5.3 kg.

VOICE
Several calls heard, somewhat similar to Adélie penguin. Contact call sharp, powerful and dissonant. Display calls include a repetitive, loud, piercingly high-pitched staccato sound (also known as 'Stonecracker' penguin as a result), a raucous cackle, soft humming and hissing. Chicks emit shrill peeping.

POPULATION AND DISTRIBUTION
Circumpolar, with large breeding colonies concentrated on islands in South Atlantic and around Antarctic Peninsula, between 54°S and 68°S. Main breeding populations found in Scotia Arc, on South Sandwich (56°18'S–59°28'S), South Orkney (60°38'S) and South Shetland (62°00'S) islands, and adjacent continental shores. Global population estimated to be at least 8 million mature individuals, slowly declining since 1980.

Dispersive, moving north to areas of broken pack-ice following seasonal sea-ice movements. Rare vagrant to Tasmania, Australia, Falkland, Crozet, Kerguelen, Marion and Macquarie Islands.

BREEDING

Extremely social, breeds from November to March in large, dense, noisy colonies usually containing tens or hundreds of thousands of birds, occasionally near Adélie and Gentoo penguins, although usually higher on rocky slopes. Colonies typically established on rough, well-drained, ice-free, elevated terrain such as rocky foreshores, headlands, outcrops and ledges up to about 75 m above sea level. Adults return from early October to November, males usually first, shortly followed by females. May start breeding from age three onwards. High fidelity to natal colony and nest site, often with long-lasting pair bonds.

Courtship: Ecstatic display similar to Adélie, but more often performed jointly, pumping chest, sky-pointing and beating flippers while emitting shrill, rasping bray.

Nest: Simple round platform with shallow bowl, made of small stones, sometimes lined with feathers and bones. More stones added during incubation and guard phase.

Laying: Normally two cream coloured eggs, similar in size, laid 2–4 days apart commencing late November-December. Occasionally one or three egg clutch. No replacement clutch if lost.

Incubation: 31–39 days. Shared by both parents in four shifts, with female usually first.

Chick rearing: If two chicks hatch, both fed equally. Shared by both parents, initially brooding/guarding for about 20–30 days. Crèches, numbering from just a few to over 100, then form with both parents foraging and provisioning, usually daily.

Fledging: 48–59 days, with first chicks departing in late February to early March.

FOOD

Mainly krill with some fish and squid, depending on location and seasonal availability, usually at depths of 45 m or less but capable of diving to about 179 m.

PRINCIPAL THREATS

Predators: At sea, mainly leopard seals. On land, Southern giant petrels, South polar skuas and gulls prey on unguarded chicks and eggs.

Fisheries: Commercial harvest of krill may represent resource competition.

Climate change: Population declines around Antarctic Peninsula have been linked to declining winter sea-ice coverage adversely affecting krill populations.

Volcanic activity: Very large colonies on volcanic islands vulnerable to severe impact especially if activity occurs during breeding and moulting season.

Disturbance/Tourism: Many colonies are too remote for human intrusion. Although limited, disturbance by tourists/scientists may increase predation opportunities around nests. 🐧

ABOVE A small colony on a snowfree rock mound, attended by scavenging Snowy sheathbills, Half Moon Island, South Shetlands.
ABOVE LEFT Pair with large chick, Paradise Bay, Antarctic Peninsula.

BELOW Most colonies are on high ground, cleared of snow by wind, Elephant Island.
BELOW LEFT Bailey Head, on volcanic Deception Island, is one of the largest colonies around the Antarctic Peninsula.

Gentoo Penguin
Pygoscelis papua

Alternative or previous names: None known.
First described: *Aptenodytes papua* J.R. Forster, 1781.
Taxonomic source: Christidis and Boles (1994, 2008); SACC (2006); Sibley and Monroe (1990); Stotz et al (1996); Turbott (1990).
Taxonomic note: Two subspecies recognised: Northern *Pygoscelis papua papua* (J.R. Forster, 1781), Subantarctic south to about 60°S; Southern *Pygoscelis papua ellsworthii* (Murphy, 1947), Antarctic Peninsula to South Sandwich Islands. In 2020, four separate species were suggested based on morphometric and genetic evidence between populations (Tyler et al. 2020).
Origin of name: '*papua*' refers to Papua New Guinea, erroneously noted source of first accounts; 'gentoo' perhaps derived from archaic word for Hindu, in reference to the white 'veil' over the head.
Conservation status: Least concern (IUCN 2020 — downlisted in 2016).
Status justification: Extremely large range and population size. Global population is assumed stable, but great fluctuations in parts of range, generally expanding southward.

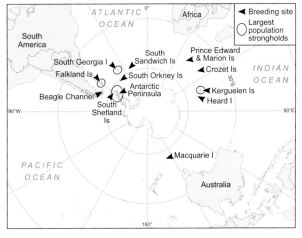

ABOVE Good nesting material, whether vegetation or pebbles depending on the region, is important to provide drainage under the nest. BELOW Pair mating. FAR RIGHT Long braying is used throughout the nesting season primarily to communicate with mates.

DESCRIPTION
Third largest penguin. Unique in appearance so cannot be confused. Fastest recorded swimming bird, with underwater speeds of up to 36 kph.

COLORATION
Adult: Black head, throat, back and tail; white ventral area. Unique white patches above eyes, narrowing to meet across crown, creating 'earmuff-like' markings. Variable white freckling over sides of head and crown, particularly in northern race. Bill mostly bright orange-red, with black on upper mandible and tip. Brown eyes with pinkish rim and variable white surround. Feet bright pink to yellow-orange.
Immature: Similar to adult but slightly smaller; less distinct colouring, with greyish back; orange bill and feet less vibrant.
> **Chick:** Pearl grey down with white undersides. Bill and feet dull orange.

SIZE AND WEIGHT
Pygoscelis papua papua slightly larger and more slender, especially longer bill than *P. p. ellsworthii*. Body length range 75–90 cm; weight variable, depending on time of year and gender, usually between 4.5–8.5 kg.

VOICE
Extremely vocal, may vary slightly depending on location. Contact call usually low-pitched and short; courtship call drawn-out staccato braying rhythmically produced while inhaling and exhaling, with a gargling raspy note. Minor disagreements involve hissing, groans accompany more serious altercations. Chicks emit varying pitched whistle and repetitive cheep when begging.

POPULATION AND DISTRIBUTION
Circumpolar, mainly on subantarctic islands and Antarctic Peninsula, between 46°S (Crozet) and 66°S (Antarctic Peninsula). 80% of global population found on Falklands (51°48'S), South Georgia (54°18'S) and South Sandwich Islands (56°18'S– 59°28'S), and the Antarctic Peninsula (including South Shetlands). Evidence of breeding population size fluctuating year by year, but overall appears stable or increasing, particularly in southern extent of range presumed in response to sea-ice reduction. Southwest Indian Ocean populations appear to be declining. Latest estimates for overall population 774 000 mature individuals (Lynch, 2013). Sedentary in subantarctic region but partially migratory farther south. Some adults present in most colonies throughout year. Usually forages within 5–25 km of colony, especially during breeding season.

Vagrant to Tasmania, New Zealand and Argentina; rarely Gough Island.

BREEDING
Colonial, from a few birds to several thousand, with some colonies relocating a few metres or even kilometres each year. Typically occupy ice-free moraines, stony coastal plains or, in Falklands, grassy slopes farther inland. Timing varies, usually commencing June to November, depending on geographical region, with prolonged season in northern subspecies increasing likelihood of second clutch if first fails. Start breeding from age 2 onwards.

High long-term pair fidelity.

Courtship: Ecstatic display performed by lone male at nest site, stretching tall and sky-pointing with loud, drawn-out call. Bonding pair performs similar mutual displays, bowing repeatedly, with low hissing sound and ritualised delivery of nest material.

Nest: Circular mound, sometimes quite large, constructed with any available materials depending on environment, including stones, mud, feathers, shells, vegetation. Theft of nest materials commonplace.

Laying: Northern range, June-November; Southern range, November-December and more synchronised. Two-egg clutch, 2–4 days apart, rarely one or three eggs.

Incubation: 35–37 days, shared by both parents, beginning with partial incubation of first egg.

Chick rearing: Both eggs usually hatch, at staggered interval reflecting laying. Parents share brooding and provisioning, alternating daily for about 20–37 days, after which highly mobile chicks form loose crèches. Returning parents often instigate long food chase, with quickest chick receiving more food, but when food supply plentiful both stand good chance of survival.

Fledging: 85–117 days in north, 62–82 days in south. Unlike other species, young seem to undergo transition period before becoming fully independent, returning to natal colony for a few weeks between gradually lengthening foraging trips to receive sporadic feeds from parents.

FOOD

Opportunistic, especially around Falkland Islands, tending to forage in waters close to breeding colony. Diet varies depending on location and season, consisting of krill, fish and squid in varying proportions usually around 100 m depth, but sometimes exceeding 200 m.

PRINCIPAL THREATS

Historically, populations affected by exploitation for commodities (oil, skin, feathers); egg harvest still permitted in some areas (under controlled licence system in Falkland Islands) but diminishing.

Predators: At sea, leopard seals, South American sea lions and orca, rarely fur seals. On land, chicks and eggs taken by Southern giant petrels, skuas, Snowy sheathbills, Kelp gulls; Striated caracara in Falklands, plus feral cats in some colonies.

Climate change: May be a factor for decline in northern part of range.

Fisheries: Prey species in all areas subject to some level of competition from commercial fisheries, with spatially restricted foraging range highlighting dependency on local resources. Actively feeding on discards from trawling fleets makes the species susceptible to bycatch (Crawford et al. 2017).

Pollution: Potential oil spills an increasing danger with expanding oil exploration around the Falklands archipelago. Contamination by ingestion of plastic particles e.g. either mistaken for food, or in turn from prey species such as Antarctic krill.

Disturbance/Tourism: More easily frightened off nest than most penguins, exposing chicks and eggs to avian predators, which could cause decreased nesting success. Associated marine traffic likely affects foraging in inshore waters.

Disease: Although not always clearly identified (e.g. recent mass mortality on Crozet & Kerguelen, C.A.Bost in litt. 2019), epidemics can significantly affect local populations, e.g. avian pox virus (2006) and paralytic shellfish poisoning from toxins caused by harmful algal bloom or 'red tide' (2002/2003), both in the Falklands. 🐧

TOP Nesting habitat varies between the Antarctic Peninsula (LEFT) and the Falkland Islands (RIGHT). ABOVE (CLOCKWISE FROM LEFT) From egg laying to fledging takes between three and five months, depending on region and food supply.
BELOW Differing features are visible between southern (left) and northern (right) subspecies.

African Penguin
Spheniscus demersus

Alternative or previous names: Black-footed penguin; Jackass penguin; Cape Penguin.
First described: *Diomedea demersa* Linnaeus, 1758.
Taxonomic source: Dowsett and Forbes-Watson (1993); Sibley and Monroe (1990).
Taxonomic note: No recognised subspecies.
Origin of name: *'demersus'* from Latin meaning 'depressed' or 'sinking'.
Conservation status: Endangered (IUCN 2020 — uplisted in 2010).
Status justification: Ongoing rapid, possibly accelerating, population decline across range, with no sign of reversal despite research and management intervention.

ABOVE Facial markings and black ventral flecks differ in every individual.

BELOW Loud braying earned the species the alternative name of Jackass penguin.
RIGHT Small groups return in the afternoon to Boulders Beach near Simonstown, South Africa, where a thriving colony that became established in 1982 is sheltered from native land predators by a coastal housing development.

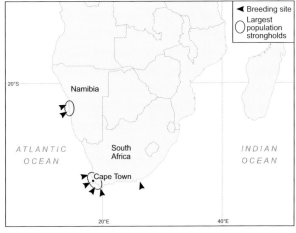

DESCRIPTION
Medium-sized, warm-climate species. The only penguin breeding on African continent.

COLORATION
Adult: Black back and tail. White ventral area, with random, individually differing black flecks. Black head with broad white stripes starting on sides of forehead, curving above eyes and around ear coverts, to join as broad white band across throat. Arching single black band across upper breast extending down each side of body, under flippers and ending in fine black line around inner thigh. Concentrated flecking sometimes gives impression of second breast band. Dark brown eyes ringed with bare pink skin which broadens above and extends towards base of upper mandible. Black bill with pale band across both mandibles near tip. Black feet with varying pink blotches.
Immature: Head and back dark blue-grey; white ventral area. Paler grey chin and throat. Banded pattern absent.
Chick: Brown head, throat and back, usually with paler face; white underparts.

SIZE AND WEIGHT
Body length 60–70 cm. Weight variable depending on gender and time of year, ranging between 2.1–3.7 kg.

VOICE
Like others in genus, a series of controlled, high-pitched inhalations and loud exhalations produces drawn-out, plaintive 'braying', heard mainly at night.

POPULATION AND DISTRIBUTION
Endemic to greater Benguela upwelling ecosystem, breeding at 28 locations in southern Namibia and South Africa, with 7 islands supporting 80% of global population. Ranges between Hollamsbird Island, off central Namibia (24°38'S) to Bird Island, Algoa Bay, South Africa (33°50'S). Largest Namibian colony on Mercury Island (25°43'S), South African on Dassen Island (33°'25'S).

Catastrophic declines, from circa 1 million pairs in early 19th century, due to guano and egg harvesting, and later oil spills and resource competition from purse-seine fisheries. Population has declined steadily in South Africa since 1979 (70 000 pairs), with a significant reduction of over 68% in just three generations (c.13 500 pairs in 2019). In the same period the Namibia population has decreased but in comparison appears to have remained more stable (c. 4300 pairs in 2019). In 2019 the global population had reduced to an historical low of approx.

17 700 breeding pairs (Sherley et al. 2020).

Mostly sedentary, remains within 40 km of colony when breeding, movements influenced by prey distribution. Some birds present in colonies all year. Foraging distances for non-breeding adults recorded between 120–350 km (Ludynia, 2007; Waller, 2011), some to 900 km. Juvenile dispersal up to 1900 km from natal colonies.

Vagrant to Gabon, Congo and Mozambique.

BREEDING

Any time of year, but some variations between locations. Small to medium colonies on islands and some mainland shores, on vegetated, flat sandy ground or steeper, bare rocky terrain. Breeding begins at ages 4–6 years. High pair bond and colony fidelity, less for nest site.

Courtship: Ecstatic display starts with sky-pointing bird throbbing silently with open bill, then braying with rhythmical flipper movement; also pairs circling, vibrating heads, body quivering, bowing, bill duelling, male patting female with flippers, and mutual preening.

Nest: Preferably burrow excavated in guano or firm sand; alternatively above ground, shaded if possible, between or under rocks/boulders, low vegetation, even buildings. May be lined with stones, shells, bones, feathers or any available vegetation. Artificial burrows provided in some locations, e.g. fibreglass igloos, pipes, and wooden nest boxes.

Laying: Synchronised within each colony, 2 white eggs up to 3 days apart. First egg larger than second. Peak laying: South Africa, March–May; Namibia, November–December. Occasionally lays second clutch, especially if first fails.

Incubation: 38–41 days, starting with first laid egg, shared by both parents with regular shift changes, usually 1–2 days.

Chick rearing: Hatching asynchronous, about 2 days apart. Brooded by both adults for 26–30 days. Small crèches then form, especially where nests exposed, with both parents provisioning.

Fledging: Highly variable, 60–130 days. Develops adult plumage in stages over 2 years.

FOOD

Forages mainly on shoaling pelagic fish, preferring anchovies, sardines and mackerel; also squid and small crustaceans, usually within 30 m depth, but capable of diving to 130 m.

PRINCIPAL THREATS

Historically killed for food, fuel for ship boilers, and rendered down for fat. Several other factors combined in drastic population declines, many still ongoing.

Predators: At sea, mainly sharks and Cape fur seals. On land, mongoose, dogs, feral cats, leopards, caracals, Kelp gulls, mole snakes and rats, depending on location.

Guano harvest: Severe, long-lasting nesting habitat degradation due to denuded burrowing substrate forcing above-ground nesting, with increased risk of overheating, flooding and predation. South Africa

harvest ceased in 1991 (~1.8 million tonnes removed as fertiliser between 1841–1983); since 2016, harvesting in Namibia no longer permitted.

Habitat: Competition with Cape fur seals and other seabirds for breeding space. Some loss due to port and coastal housing developments.

Egg harvesting: Historically very serious. Although illegal since 1967, recorded poaching incident in South Africa in 2016.

Fisheries: Pressure on prey species by commercial purse-seine fisheries considered significant factor in breeding failures; exacerbating competition with Cape fur seals. Gill netting near colonies may increase entanglement mortality. Bearded goby alternative prey species in Namibia, but energy-poor so inadequate for breeders.

Oil pollution: Severe ongoing impact from accidental and intentional oil spills (bilge-cleaning, ship-to-ship bunkering, port operations and developments close to colonies). Since 1990 number of oiled birds has increased significantly and, despite rescue, clean-up and rehabilitation efforts, appears to have long-term impact on colonies and breeding success rates.

Disease: Records show species affected by a number of diseases (some non-lethal). Few mass mortality events until highly pathogenic avian flu killed 100 South African and approx. 600 Namibian penguins in 2018/19.

Climate change: Warming sea temperatures apparent factor in changing prey distribution. Eastward trend of sardine and anchovy schools affect reproduction due to excessive distance from South African nesting sites (see Ryan, p. 182).

Disturbance/Tourism: Mainland colonies susceptible to disturbance and road traffic.

ABOVE Historically, Robben Island, just outside Cape Town (in background), was a major wildlife haven ('robben' means fur seal in Afrikaans), but like many has suffered from prolonged human presence — its penguins had been exterminated by the 1800s. Now a Cultural World Heritage Site (ex-penitentiary buildings shown above), conservation efforts mean it is again home to 3600 pairs of African penguins — the third largest colony in South Africa — though it is still affected by oil spills. ABOVE LEFT Immatures differ from adults in having no contrasting face and chest bands, which are acquired gradually over two years. BELOW Parent and chick preening at the Boulders Beach colony, South Africa.

Magellanic Penguin
Spheniscus magellanicus

Alternative or previous names: Patagonian penguin; Jackass (Falkland Islands).
First described: *Aptenodytes magellanicus* J.R. Forster, 1781.
Taxonomic source: Christidis and Boles (1994, 2008); SACC (2006); Sibley and Monroe (1990); Stotz et al (1996); Turbott (1990).
Taxonomic note: Some consider *S. magellanicus* to be conspecific with *S. humboldti* and *S. demersus*.
Origin of name: *'magellanicus'* after 16th-century explorer Ferdinand Magellan, who saw them on his initial voyage around the tip of South America.
Conservation status: Least concern (IUCN 2020 — downlisted in 2020).
Status justification: Fluctuating numbers in different parts of range, but overall considered stable or slowly declining.

ABOVE Pair guarding nesting burrow entrance, Cabo Dos Bahias, Argentine Patagonia. BELOW, CENTRE AND FAR RIGHT Courtship involves exaggerated posturing, strutting, flipper tapping and braying.

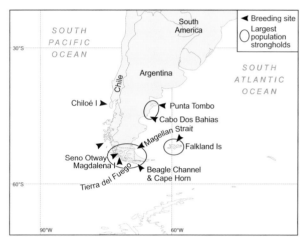

Chick: Greyish-brown head and back with white underparts. Pale cheeks and chin.

SIZE AND WEIGHT
Body length 70 cm; weight variable, depending on time of year and gender, usually between 2.3–7.8 kg.

VOICE
Like others in genus, a series of controlled, inhalations and loud exhalations culminating with drawn-out, plaintive 'braying', heard mostly early morning and evening. Used in courtship, also before and after a disagreement. Once paired, generally silent during rest of breeding season except for greetings, during fights and nest changeovers. Chicks emit simple 'cheep'.

DESCRIPTION
Largest of the warm-climate 'banded' penguins. Extremely feisty and aggressive. Could be confused with Humboldt penguin, only similar species with slight range overlap; easily distinguishable on land, having two black bands rather than one.

COLORATION
Adult: Black back and tail. White ventral area, with a few random, individually differing black flecks. Black head with white stripes starting on sides of forehead, curving above eyes and around ear coverts, to join as white band across throat. In frontal view, two black bands: one broad, crossing just beneath white throat band, joining black dorsal area at shoulders; second slightly narrower band arches across upper breast then extends down each side of body, under flippers and ending in finer black line around inner thigh. Brown eyes ringed pink. Bare pink skin from above eyes extending towards upper mandible, contrasting with black skin around gape. Black bill with variable pale marking near tip, mainly lower mandible. Black feet with varying pink blotches.
Immature: Smaller, grey dorsal plumage. Variable pale grey to white cheeks and chin. No banding.

POPULATION AND DISTRIBUTION
Breeds around Falkland Islands and both mainland and islands along Patagonian coast of Argentina and Chile. Breeding range on both seaboards extends southwards from approximately 41°S (Golfo San Matías, Argentina; northern Chiloé Island, Chile) to Tierra del Fuego and Cape Horn (55°58'S). Largest colony at Punta Tombo (44°02'S) in Argentina. Population trends vary across range, some colonies showing substantial declines while others are stable or increasing. Global population estimates vary between 1.1 and 1.7 million pairs, with the majority breeding in Argentina.

Population trends vary across range, some colonies showing substantial declines. Global population estimated 1.3 million pairs, about 73% breeding in Argentina.

Migratory. Absent from colonies during pre-moult period and May-August, Atlantic and Pacific birds travelling northwards as far as Brazil and Peru respectively. Foraging distances

during breeding season highly variable, both between colonies and individuals. Examples: 14–120 km from colony, up to 75 km offshore, round-trips up to 283 km travelled, 8–74 hours duration. (Sala et al, 2012).

Vagrant to Antarctic Peninsula, subantarctic islands, New Zealand and Australia.

BREEDING

Extremely social in continental colonies; loosely colonial in Falkland Islands. Slight variation in season depending on location. Nesting habitat varies: level or tiered terrain, dunes, tussock grass, shrubs, forest, up to 70 m above sea level and 1 km inland. May be found in proximity with Gentoo and/or King penguins in Falkland Islands and Beagle Channel. Range overlap with Humboldt penguin in Chile (3 colonies known). First documented Humboldt-Magellanic hybrid, Puñihuil Island (Simeone 2009).

Begins breeding from about 4 years but more often age 5–8. Males return to colonies in September, females shortly afterwards. High pair bond and nest site fidelity.

Courtship: Commences with ecstatic display usually by male standing at nest site, stretching tall and pointing open bill skywards, rhythmically moving outstretched flippers and emitting repetitive low-pitched 'huff' which crescendoes into loud bray. Other displays involve bill fencing, male circling female, male tapping flippers against female back and sides, and mutual preening.

Nest: Deep burrow dug in various substrates, from hard glacial deposit to sand; sometimes shallow depression/scrape preferably sheltered under bushes, tree roots or rocks; may be lined with sticks, grass or any other available materials.

Laying: Synchronous within colony. 2 off-white eggs, about 4 days apart, commencing early to mid-October. First laid egg larger than second.

Incubation: 39–42 days. Shared, female for first of two approximately 15–day shifts, followed by shorter shifts until hatching.

Chick rearing: Both eggs usually hatch, about one day apart. Parents initially alternate brooding and foraging for about 30 days; chicks later left unattended while both forage. No crèches formed, chicks remain in/near burrow until fledging. Adults favour first hatched when provisioning, so younger chick may starve during times of food shortage; both raised in ideal conditions.

Fledging: Anywhere between 60–120 days, January to early March (food supply dependent).

FOOD

Depending on location, mainly anchovy and sardine,

some cephalopods and crustaceans, usually at depths averaging 30 m, with deepest dive recorded 91 m.

PRINCIPAL THREATS

In 1981, plans for commercial exploitation at Punta Tombo for leather, oil and protein halted when Wildlife Conservation Society (New York) teamed up with Chubut Province Tourism Office, landowners and Dr P. Dee Boersma (see p. 184) to undertake long-term study.

Predators: At sea and on beaches, South American sea lion and giant petrels. On land, caracaras, skuas and gulls take exposed chicks and eggs. Some areas with tourism show less predation. Depending on location, foxes, feral cats and dogs. Exploited by humans at some sites in Chile, for food and bait.

Fisheries: Resource competition thought to be contributing factor in some population declines. Overfishing of prey species, especially anchovy and squid, forces parents to forage at greater distances causing stress in chick provisioning. Accidental entanglement/drowning in gillnets.

Habitat: Overgrazing by domestic livestock causing collapse of burrows.

Egg harvesting: Still permitted in Falkland Islands (under licence) but diminishing.

Climate change: Danger of increased rainfall causing flooding/collapse of burrows, conversely drought killing sheltering vegetation or making ground unsuitable for burrowing.

Oil pollution: Constant risk from offshore oil wells and tanker traffic in Argentina and Chile, plus expanding oil exploration around Falklands archipelago.

TOP LEFT Striking colour differences between adults and immatures.
TOP RIGHT Group courtship is frequent, Cabo Dos Bahias, Argentina.
ABOVE LEFT Family greeting, Saunders Island, Falklands.
ABOVE Landing in surf, Punta Tombo, Argentine Patagonia. Facial markings are conspicuous even at sea.

Humboldt Penguin
Spheniscus humboldti

Alternative or previous names: Peruvian penguin, Patranca or Pájaro niño.
First described: Meyen, 1834.
Taxonomic source: SACC (2006); Sibley and Monroe (1990); Stotz et al (1996).
Taxonomic note: No recognised subspecies.
Origin of name: '*humboldti*' after the cold water current to which it is restricted, in turn named after Prussian naturalist and explorer Alexander von Humboldt, 1769–1859.
Conservation status: Vulnerable (IUCN 2020 — uplisted in 2000).
Status justification: Extreme population declines in Peru and fluctuations in Chile, plus loss of breeding sites; probable continuing decline.

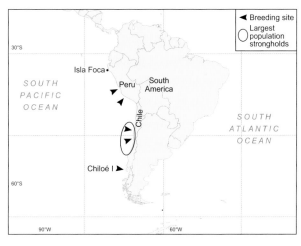

BELOW (BOTH) With natural hollows in short supply, some nests are exposed, risking predation from turkey vultures or overheating on sunny days.

DESCRIPTION
Medium-sized, temperate-region penguin. Can be confused with Magellanic penguin, only similar species with range overlap. Easily distinguishable on land, having only one black band rather than two. Most common zoo penguin worldwide.

COLORATION
Adult: Blackish back and tail. Ventral area white, with random, individually differing black flecks. Head mostly black with slender white stripe starting above each eye, running in gradually widening crescent around ear coverts, to join as a broad band across lower throat. Single black chest band extending down each side of body, under flippers and ending in fine black line around inner thigh. Reddish brown eyes often with pink rim that joins pink fleshy area at base of bill, extending beneath chin. Black bill, with faded peach mainly at base of lower mandible and pale band across both mandibles towards tip. Feet black with varying pink patches.
Immature: Mostly dark grey body and head (paler on sides of head and chin), light underside; no breast band.
Chick: Dusty grey, with white belly, cheeks and throat, but dark around base of bill.

SIZE AND WEIGHT
Body length 65–70 cm; weight variable, depending on time of year and gender, usually between 4–5 kg.

VOICE
Like others in genus, drawn-out, plaintive 'braying', somewhat less forceful than Magellanic, performed mostly at night.

POPULATION AND DISTRIBUTION
Endemic to temperate coasts and islands of Peru and Chile, in areas nourished by Humboldt Current upwelling. Range extends from Isla Foca (5°12'S) in northern Peru to Isla Guafo (43°32'S) south of Chiloé Island, Chile, where it overlaps with the range of the Magellanic penguin. 47 known breeding sites, with largest colonies at Punta San Juan, Peru and Isla Chañaral, Chile.

Thought to have exceeded 1 million in mid 19th century, but declined severely between 1880 and 1930s due to guano harvesting for fertiliser, and later resource competition by fishmeal industry. Population data from last 40 years varies considerably. 2017 counts in both Peru and Chile suggest an overall maximum population of 23 800 mature individuals, though seasonal timing and data collection methodology differ across the range, which makes interpreting population trends more difficult, and requires further research.

Originally thought sedentary, normally forages within 35–50 km of nesting site during breeding season, but evidence of larger distances covered in response to environmental factors and prey shortages, with non-breeding dispersal in Peru up to 170 km, less frequently 600 km. New research indicates possibility of approximate 700 km migration between Peru and northern Chile.

Vagrants recorded as far north as Colombia; individuals seen in Alaskan waters, but not known whether naturally or by human transport.

BREEDING

Small colonies on offshore islands, occasionally mainland coast. No fixed season, but mostly during southern winter. In ideal conditions, may rear two successive clutches of two chicks each, although youngest may starve during times of food shortage. Begins breeding at about 4 years old. Monogamous, although extra-pair copulation observed; high fidelity to breeding site.

Courtship: Commences with ecstatic display usually by male at nest site, pointing open bill skywards and emitting repetitive 'donkey-like' bray, with slowly flapping flippers; performed jointly in bonding pairs, with more forward pointing bills. Other rituals involve male tapping flippers against female back and sides, bill fencing, bowing and mutual preening.

Nest: Burrows excavated in ancient guano or hard soil; may utilise natural crevices and caves or scrapes under overhang or desert vegetation. Often lined with feathers or seaweed.

Laying: Two similarly sized eggs, 2–4 days apart. Peru: anytime between March and December, with peaks April and August-September. Chile: generally one month later.

Incubation: 40–42 days. Shared by both parents, with regular shift changes.

Chick rearing: Both eggs usually hatch, 2–4 days apart. Chicks stay in burrow/nest for 2–3 weeks, then remain nearby until fledging. No crèches formed. Parents initially alternate between brooding and foraging; chicks later left unattended while both forage.

Fledging: 70–90 days, departing to spend several months at sea. Moult to adult plumage at about one year.

FOOD

Feeds mainly in inshore waters on anchovies, aracaurian herring, silverside, garfish, crustaceans and cephalopods, normally at depths of less than 60 m.

PRINCIPAL THREATS

Predators: Orca and fur seals at sea, though not known as significant. On land: wild dogs, desert foxes, kelp gulls and rarely caracaras and turkey vultures. In some colonies, feral dogs and cats, and in particular rats.

Fisheries: Industrial, large-scale trawl, purse-seine and longline fisheries exploit main prey species, with penguins as common bycatch. Widespread artisanal gillnet fisheries pose a particular threat from entanglement throughout range, potentially with greater negative impact than large-scale fisheries (Crawford et al. 2017), especially during the winter dispersal away from colonies. Blast fishing with explosives, primarily in Peru, can also cause penguin mortality.

Guano harvest: Practised since Inca times, but reached devastating scale during last quarter of 19th century, and continues to affect breeding success. Today Peruvian government institute manages guano extraction, providing legal protection at 12 principal colonies, with walls (to restrict access by intruders and small carnivores, as well as prevent erosion in some places) and guards posted at some sites.

Mining: In Chile, successful campaign prevented con-struction of coal mine at Punta Choros, although threat remains of two coal-fired power stations being built.

Hunting: Coastal fishermen and guano miners (Peru) may consume adults and chicks, or use them for bait. In Chile, a 30-year hunting moratorium commenced in 1995.

Climate change/El Niño Southern Oscillation (ENSO): Huge population declines occur following ENSO events. Nests can flood with high sea levels, plus sharp rise in sea temperature and suppressed nutrient-rich upwelling responsible for mass starvation and breeding failure. Climate models predict intensified ENSO events could spell problematic future.

Oil pollution: Some colonies could be affected by oil spills, as happened with two major spills near the Cachagua colony in central Chile in 2015/2016.

Disturbance/Tourism: Extremely sensitive to human presence; appears incompatible even with controlled eco-tourism, with reduced breeding success demonstrated in 2006 Chilean study on Damas, Choros and Chañaral islands. If an approved industrial mega-port proceeds, it could threaten the largest colony at Punta San Juan, Peru.

ABOVE LEFT Lacking ancient guano layer in which to dig burrows, nests are sited under natural shelters, either boulders or scant vegetation, all photos from Tilgo Island, northern Chile.
ABOVE Nesting under collapsed salt bush, overlooking typical desert coast.

BELOW Groups coming ashore in late afternoon are nervous and easily frightened, one of the few species not suited to ecotourism.
FAR LEFT The species is limited to the nutrient rich waters of the Humboldt Current.

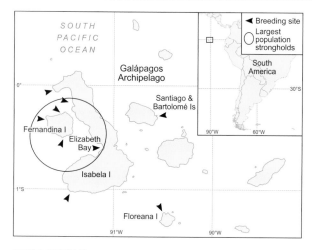

Galapagos Penguin
Spheniscus mendiculus

Alternative or previous names: None known.
First described: Sundevall, 1871.
Taxonomic source: SACC (2006); Sibley and Monroe (1990); Stotz et al (1996).
Taxonomic note: Rarest penguin with small population gene pool and highly restricted range.
Origin of name: Derived from latin *'mendicus'* meaning 'beggarly, paltry, pitiful', *'mendiculus'* refers to small size of species. 'Galapagos' endemic to the Galapagos Islands.
Conservation status: Endangered (IUCN 2020 — uplisted in 2000).
Status justification: Extremely susceptible to natural events or human impact due to tiny population and very restricted range, with majority breeding on just one island. Long-term monitoring indicates population subject to severe fluctuations, with significant and rapid decline in last three decades, mainly due to climatic disturbances apparently becoming more acute.

FAR RIGHT The equatorial, volcanic habitat of the second smallest of all penguins is also the most unusual, Elizabeth Bay, Isabela Island.

BELOW The scar from a shark bite is visible in the feathers of this otherwise healthy bird, Cape Douglas, Fernandina Island.

DESCRIPTION
Smallest of four 'banded' penguins and northernmost of all species, breeding in tropical climate, but surrounded by cooler waters. Most closely related to Humboldt penguin. Confusion with other species impossible due to diminutive size and complete range segregation.

COLORATION
Adult: Black back and tail. Ventral area white, with random, individually differing black flecks. Head and throat mostly black with variably white chin. Thin white stripe starting at corner of each eye and looping down around ear coverts to meet at throat. Black breast band, with corresponding white line arching across upper breast and extending down each side of body, under flippers and ending at outer thighs. These bands less well defined than in other Spheniscus penguins, sometimes merging together, plus varying amount of interspersed black and white feathers in lower area creates freckled, uneven appearance. Black upper mandible, lower mandible yellowish-pink with black tip; pink fleshy area at base of bill extending to chin and speckled with black during breeding season. Brown eyes with bare pink skin surround, also speckled black. Black feet with whitish blotches

and black soles. Sexes similar, but males more boldly marked with more white on chin and larger unfeathered pink area around bill.
Immature: Dark grey body and head with whitish cheeks but no white chin. No breast bands and no unfeathered area on head. Pinkish eyes. Dark bill.
Chick: Grey-brown head, throat and back with white underparts.

SIZE AND WEIGHT
Body length 48–53 cm; weight variable, depending on time of year and gender, usually between 1.4–2.9 kg.

VOICE
Donkey-like braying typical of genus, but much softer and more 'mournful' sounding than larger relatives. Likewise 'social' courting or greeting calls, involving two or more individuals, are low volume, heard mostly at night. Contact call a gentle honk, reminiscent of short blast of distant foghorn.

POPULATION AND DISTRIBUTION
Endemic to equatorial Galapagos Islands. Small, restricted range, preferring western islands where waters are coolest. About 95% of population occurs around Fernandina (0°22'S) and west coast of Isabela (0°30'S) islands. Remaining population located in small

pockets on Bartolome (0°17'S), Santiago (0°15'S) and Floreana (1°17'S) islands. Last complete census undertaken in 2009, with total number counted 1042 individuals (see Vargas, p. 180).

Normally sedentary, foraging in very shallow coastal waters, usually within 200 m from shore and a few kilometres coastal range (maximum recorded 23.5 km) from colony. Non-breeders, mostly juveniles, occasionally visit adjacent areas: southern coast of Isabela, Sombrero Chino, Rábida, Pinzón, western Santa Cruz, rarely reaching San Cristóbal, Española, Seymour and Baltra islands during severe El Niño events, when prey species fail.

Recorded as vagrant to Panama but most likely transported by ship.

BREEDING

Loosely colonial nesting in very small numbers. May breed any time of year, except when moulting, with potential to raise more than one clutch, provided water temperature is less than 24°C to ensure optimum availability and quality of prey. Mostly active at twilight or nocturnal at breeding sites. Commences breeding at approximately 4 years old. High pair bond and nest site fidelity year to year.

Courtship: Performed mostly in twilight or at night in pairs or small groups. Begins with lone males braying to attract attention, then group strutting, braying and 'chortling', posturing rigidly with jerking head movements and throat puffed out. Pairs continue with bill duelling, mutual preening and flipper patting as an invitation to mating.

Nest: Natural cavities in lava crevices or volcanic rubble, underneath boulders or tuff and in small lava tubes providing shelter from equatorial sun. Very basic nest lined with twigs, feathers and available vegetation.

Laying: Two-egg clutch, with 2–4 day interval. Climate dependent, but usually June–September, or December–March in La Niña years, less frequent in April–May.

Incubation: Shared equally by both parents for 38–42 days, starting when first egg laid.

Chick rearing: Both eggs usually hatch. Chicks brooded for about 30 days by both parents with regular shift changes. Chicks remain at or near nest site and do not form crèches.

Fledging: 60–65 days.

FOOD

Feeds in coastal waters mainly on schooling sardines, anchovy, piquitangas and small mullet, diving at shallow depths, usually less than 6 m, with maximum dive recorded 52.1 m.

PRINCIPAL THREATS

Predators: At sea, mainly by sharks. On land, Galapagos hawks, feral cats and dogs, where present, prey on adults and juveniles. Snakes, crabs, owls and rats may take unprotected young chicks and eggs.

Climate change: Increasing frequency, severity and duration of El Niño events, causing rise in sea temperatures and concurrent dispersal of prey, threatens long term survival of species, (see Vargas, p.180).

Fisheries: Accidental bycatch in coastal fisheries, especially drowning in illegal gillnets, prohibited in Galapagos Marine Reserve.

Disease: Real and potential effects of pathogens and parasites introduced to islands by human activities, for example, *Plasmodium* type parasite found in penguin blood samples taken 2003–2006, though no immediate effects on health noted at that time (see Vargas, p. 180).

Volcanic activity: With main breeding habitat centred around highly active volcanoes, large numbers potentially at risk from lava flows reaching coast.

Oil pollution: Spills from fuel tankers or grounded tour ships could affect major parts of highly restricted breeding and feeding range, with potential to impact overall population. ⬆

ABOVE Adults and immatures rest on the lava shoreline of southern Isabela Island, where nest predation by cats has been identified as an ongoing management concern for the Galapagos National Park.
ABOVE CENTRE Typical habitat shared with sally-lightfoot crabs.
ABOVE LEFT (BOTH) Generally restricted to the Cromwell Current upwelling around the western islands, the species feeds in coastal shallows mostly within 200 m of the shore.

BELOW Colour difference between adult (right) and immature.

Southern Rockhopper Penguin
Eudyptes chrysocome

Alternative or previous names: None known, though informally nicknamed 'Rockies' in Falklands.
First described: *Aptenodytes chrysocome* J. R. Forster, 1781.
Taxonomic source: Christidis and Boles (2008).
Taxonomic note: Northern *Eudyptes moseleyi* separated from Southern *E. chrysocome* in 2006. The latter has two recognised subspecies: *E. chrysocome chrysocome* (Western rockhopper) and *E.c filholi* (Eastern rockhopper). Based on recent genetic studies, some scholars suggest that *E. c. chrysocome* be further distinguished as two distinct subpopulations (Lois *et al.* 2020).
Origin of name: *'chryso'* from Greek *'khrusos'* meaning 'gold', *'come'* from Ancient Greek *'kom'* meaning 'hair'.
Conservation status: Vulnerable (IUCN 2020 — since 2008).
Status justification: Significant declines over a long period, apparently ongoing and projected to continue over the next three generations.

RIGHT (CENTRE) Photo sequence showing remarkable jumping ability, using springy leg muscles to absorb landing impact, Deaths Head, West Falkland.

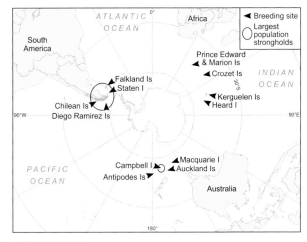

BELOW Immature coloration, with bluish-black back, dark bill and barely visible crest plumes, Campbell Island, New Zealand Subantarctic.

DESCRIPTION
Extremely gregarious and aggressive, breeds and moults in tight aggregations. Similar to Northern rockhopper penguin but smaller, with shorter flippers and bill, less black on underside of flippers and much less flamboyant crest. May be confused with other crested penguins, especially juveniles or when at sea.

COLORATION
Adult: Greyish-black head, throat, back and tail. White ventral area. Narrow, yellow superciliary (eyebrow) stripes start near base of bill, passing above eyes towards back of head where longer, yellow plumes project outwards or droop down; merge with pronounced bristling black crest across back of crown. Bright red eyes. Sturdy, orange-red bill; Eastern rockhopper has variable pale pink lower mandible; bare pink skin around bill, becoming triangular at gape. Pink feet with black soles.
Immature: Grey chin and throat with some white speckling; minimal or indistinct superciliary stripe. Dark brown bill and eyes.
Chick: Grey-brown head and back with white underparts.

SIZE AND WEIGHT
Body length 45–58 cm. Weight variable depending on gender and time of year, ranging between 2.0–3.8 kg.

VOICE
Extremely loud, harsh and shrill (has been likened by some authors to sound of a 'rusty wheelbarrow'). Contact call is short and sharp. In courtship, includes strident and extended braying and growl-like guttural throbbing, also used in greetings between mates and neighbours. Disagreements include sharp screams, high-pitched guttural throbs and grunts. Chicks emit simple cheeps.

POPULATION AND DISTRIBUTION
Circumpolar on islands north of polar front, in southern Atlantic, Indian and southeastern Pacific oceans. Nominate Southern rockhopper breeds in Falkland Islands and several offshore islands of southern Argentina and Chile. Wider ranging Eastern rockhopper subspecies breeds on subantarctic Antipodes (49°41'S), Auckland (50°42'S), Campbell (52°32'S), Macquarie (54°37'S), Heard (53°'06'S), Kerguelen (49°21'S), Crozet (46°25'S), Marion (46°54'S) and Prince Edward (46°38'S) Islands.

Huge declines across much of range since early 20th century, thought to be in many millions. Global population now estimated at 1.27 million breeding pairs representing 34% overall decline (although calculations partly based on 1980s figures), with much greater localised declines, e.g. 94% over 40 years on Campbell Island (see Thompson p. 186).

Dispersive. Absent from colonies during pre-moult period and from April–May to October. Distances travelled to foraging grounds vary depending on location but capable of long trips, some recorded to 2000 km. During breeding season foraging sometimes in excess of 100 km.

Vagrant to South Africa, New Zealand, Australia and rarely South Georgia.

BREEDING

Extremely social. Forms dense, noisy colonies (usually many thousands of pairs), on rough bouldery ground, ranging from level to very steep, either bare rock or under dense tussock grass, even in rock crevices depending on location. Access to colonies mostly precipitous, up to 60 m above sea level and several hundred metres inland. Often sharing space with Black-browed albatross and Imperial cormorants. Season variable depending on region, northern part of range earlier than south, which breed October to March. Probably commence breeding at about age 4–5 years. Relatively high pair bond and nest site fidelity.

Courtship: Male performs ecstatic display, swinging head vertically and widely from side to side, with loud, throbbing brays and pumping flipper movements. Also includes excited hopping around, mutual braying, bowing, quivering and preening.

Nest: Small shallow depression lined with any available debris, e.g. feathers, tussock, mud, stones, bones or other items. Will also use abandoned pedestal nests of Black-browed albatross.

Laying: Highly synchronous within colonies. Two egg clutch, 4–5 days apart, from early November. First A-egg smaller than second B-egg.

Incubation: 32–34 days. Shared by both adults in 3 roughly equal shifts, first remaining on site together, then male departing to forage, trading place with female for final 2 weeks before hatching.

Chick rearing: Hatching normally from B-egg, in early December. A-egg may hatch (a day later) but chick usually dies from neglect and starvation within first week (See Morrison, p. 174). Falkland Islands birds sometimes raise 2 chicks if ideal conditions prevail. Brooded by male for 24–26 days, female alone foraging and feeding chick. Small crèches then form, with initial provisioning by female, then by both parents.

Fledging: 66–73 days, departing early February. Develops adult plumage at age 2 years.

FOOD

Varies depending on season and location, mainly krill, squid, octopus, fish and other small crustaceans in varying proportions, with dive depths averaging 15–45 m.

PRINCIPAL THREATS

Historically, populations decimated by exploitation for oil extraction, plus eggs harvested for food. Egging prohibited in Falkland Islands since 1999.

Predators: At sea, include sharks, South American sea lion, fur seals, giant petrels. On land, mainly skuas, striated caracaras and gulls prey on chicks and eggs. Poorly protected sites in Chile still threatened by human predation mostly for baiting crab pots.

Fisheries: Overfishing in some parts of range suggested as factor contributing to population declines.

Habitat loss: Degradation by grazing livestock in some parts of range. Before eradication, rabbits caused erosion on Macquarie.

Climate change: Several studies suggest link between population declines and reduction of prey availability with warming sea temperatures.

Oil pollution: Threat from hydrocarbon exploitation off Argentina's Patagonian coast and expanding oil exploration around Falklands archipelago.

Disease: Although not always identified, epidemics can significantly affect local populations, e.g. avian cholera on Campbell Island (1985/86) and paralytic shellfish poisoning from toxins caused by a harmful algal bloom or 'red tide' (2002/2003) in the Falklands.

ABOVE Many colonies are mixed with Black-browed albatross, Saunders Island, Falklands.
ABOVE CENTRE Pair greeting at nest, New Island, Falklands.

BELOW When feeding conditions are exceptionally favourable, two chicks may be raised, Hope Harbour, West Falkland.
BELOW LEFT (BOTH) Colour difference between western (Falkland Islands, LEFT) and eastern (Campbell Island, RIGHT) subspecies are clearly visible, especially around bill and gape.

Northern Rockhopper Penguin
Eudyptes moseleyi

Alternative or previous names: Known as Pinnamin by Tristan Islanders; Long-crested rockhopper penguin; Moseley's penguin.
First described: Mathews and Iredale, 1921.
Taxonomic source: Banks, Van Buren, Cherel and Whitfield (2006).
Taxonomic note: In 2006, following research which demonstrated morphological, vocal and genetic differences, the Rockhopper penguin *Eudyptes chrysocome* was split into two separate species with northern *E. moseleyi* elevated to species level, formerly a subspecies of southern *E. chrysocome* (Jouventin *et al*, 2006).
Origin of name: '*moseleyi*' after Henry Nottidge Moseley, a British naturalist aboard HMS *Challenger* during 1872–1876 global scientific expedition.
Conservation status: Endangered (IUCN 2020 — since 2008).
Status justification: Limited range in temperate South Atlantic and Indian oceans. Evidence of rapid population decline in last three decades, but reasons largely unknown.

ABOVE RIGHT The crest in this species is by far the most extravagant in the genus, Gough Island.
RIGHT Wherever possible, nests are tucked out of sight within tall, bamboo-like tussock grass thickets, Gough Island, South Atlantic.

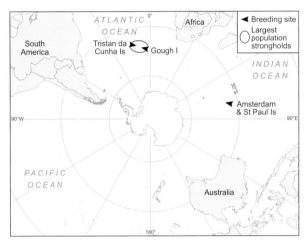

DESCRIPTION
Similar to Southern rockhopper penguin but larger, with longer flippers and bill, more black on underside of flippers and much more flamboyant crest. Also, breeding season earlier and chick-rearing shorter. Oceanic forager during breeding season, unique amongst Eudyptids, as breeds on volcanic islands lacking coastal shelves (Tremblay et al, 1997).

COLORATION
Adult: Greyish-black head, throat, back and tail. White ventral area. Narrow, yellow superciliary stripes start near base of bill, passing above eyes towards back of head, becoming very long, abundant yellow and black plumes which project outwards or droop loosely down, merging with short black crest across back of crown. Bright red eyes. Sturdy, orange-red bill. Pink feet with black soles.
Immature: Similar to adult, but slightly smaller and slimmer. Grey chin and throat, speckled white. Eyes and bill much duller colour. Crest very short and black, lacking long plumes; yellow superciliary stripes hardly noticeable, or absent.
Chick: Grey-brown head, throat and back with off-white underparts.

BELOW Immatures show a black crest before the exaggerated yellow plumes develop.

SIZE AND WEIGHT
Body length 55–65 cm. Weight variable depending on gender and time of year, ranging between 1.6–4.0 kg.

VOICE
Similar to Southern rockhopper penguin but with deeper, slightly hoarser tone.

POPULATION AND DISTRIBUTION
Breeding restricted to Tristan da Cunha (37°07'S) and Gough Island (40°19'S) group in South Atlantic, accounting for approx. 90% of global population. Remainder breeds on Île Amsterdam (37°50'S) and Île St Paul (38°43'S) in Indian Ocean. Largest colony on Middle Island, Tristan da Cunha group.

Once numbering in their millions, species has suffered massive declines since 1950s, over 1 million disappearing from Tristan da Cunha and Gough Island, latter showing 98% decrease between 1955 and 2006. Population counts difficult due to remote locations, but last published figures estimate global population around 206 850 breeding pairs.

Migratory. Absent from colonies during pre-moult period and April–June; little known about movements.

Vagrant to South Africa, South America and rarely Falkland Islands.

BREEDING

Extremely social, moults and breeds in dense, noisy colonies, usually with thousands of pairs, established near coast under dense tussock grass, or on rocky slopes, in crevices or small caves usually accessed via steep tracks, often formed by erosion. Breeding season slightly variable depending on location, late July to early January on Tristan da Cunha group; 3–4 weeks later on Gough Island. High pair-bond fidelity.

Courtship: Similar to Southern rockhopper; display involves raising flippers, throwing head back, shaking head plumes and braying loudly.

Nest: Small shallow scrape lined with feathers, tussock, stones or other available materials.

Laying: 2-egg clutch, 4–5 days apart, early September. First A-egg smaller than second B-egg.

Incubation: 32–34 days. Both birds remain at nest sharing incubation for first 12 days, then two alternate shifts, female about 12 days, then male until hatching.

Chick rearing: Hatching mid-October normally from B-egg. A-egg may hatch (about a day later) but chick usually dies from neglect and starvation within first week (See Morrison, p. 174). Brooded by male for first 20–26 days, with female alone foraging and feeding chick. Loose crèches then form, with initial provisioning by female, then by both parents.

Fledging: End December–early January.

FOOD

Varies depending on season and location, mainly krill, squid, octopus, fish and other small crustaceans in varying proportions, with average dive depths less than 20 m, rarely over 90 m but capable of diving over 100 m.

PRINCIPAL THREATS

Range reduction on Tristan da Cunha in part caused by historic exploitation by inhabitants for food, bait and decorative head plumes. Feral pigs posed huge problem prior to eradication (Tristan in 1873, Inaccessible in 1930).

Predators: At sea, sharks, subantarctic fur seals (especially at Amsterdam and Gough Islands) and Northern giant petrels (Ryan et al, 2008). On land, abundant skuas nesting nearby prey on chicks and eggs. On Gough Island, introduced house mouse preys on chicks and eggs of albatrosses and petrels, posing potential threat until eradicated.

Fisheries: Historically, significant mortality in crayfish fisheries, penguins used as bait, now ended.

Habitat: Competition for nesting space with increasing populations of subantarctic fur seals. On main Tristan Island, tussock grass initially cleared for growing crops. Degradation by grazing livestock and erosion.

Egg harvesting: Continues under Conservation of Native Organisms and Natural Habitats (Tristan da Cunha) Ordinance 2006, allowing harvest for domestic purposes from Nightingale, Middle and Stoltenhoff Islands.

Climate change: Increases in sea temperatures and resultant changes in distribution of prey species.

Oil pollution: Though not in direct regions of drilling or tanker activity, shipping accidents can severely impact an already declining species, as experienced in 2011 grounding of the cargo vessel *Oliva* at Nightingale Island, heavy fuel oil also reaching Inaccessible and Tristan da Cunha. 3718 oiled penguins were collected and taken to main island for rehabilitation, with only 381 surviving to release. Although 1000 unoiled penguins were also saved it is likely many more perished in out-of-reach areas. (See Glass, p. 188). 🐧

ABOVE Part of the Tristan da Cunha group, Inaccessible Island is a pristine World Heritage site with several healthy nesting colonies.
ABOVE LEFT Subantarctic fur seals invade the colonies on Gough Island, sometimes crushing eggs and chicks.

BELOW With small chicks in the colony, large numbers of adults head out to feed at sunrise, Nightingale Island, Tristan da Cunha.
BELOW LEFT A pair tends to their small chick on Gough Island.

Macaroni Penguin
Eudyptes chrysolophus

Alternative or previous names: None known.
First described: *Catarrhactes chrysopholus* Brandt, 1837.
Taxonomic source: Christidis and Boles (1994, 2008); SACC (2006); Dowsett & Forbes-Watson (1993); Sibley and Monroe (1990); Stotz et al (1996); Turbott (1990).
Taxonomic note: Some consider Royal penguin *E. schlegeli* to be subspecies, but most authors treat them separately.
Origin of name: *'chrysopholus'* derived from Greek meaning 'gold crest or tuft'. 'Macaroni' was a term generally used to describe an 18th-century British dandy, with exaggerated hairstyles and mannerisms.
Conservation status: Vulnerable (IUCN 2020 — uplisted in 2000).
Status justification: Apparent rapid declines over a 37-year period (three generations).

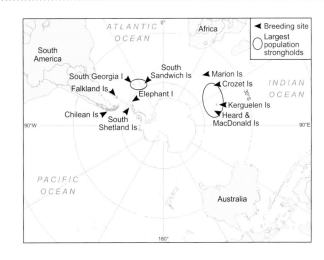

BELOW Pairs stand guard together, taking turns on the nest during the first stage of incubation, Candlemas Island, South Sandwich group.

DESCRIPTION
The most abundant of all penguin species, and most southerly ranging among seven 'crested' penguins. Confusion with closely related Royal penguin unlikely due to black vs. white face, but distinction less clear in grey-throated juveniles of both. May be hard to differentiate from other crested species at sea.

COLORATION
Adult: Blue-black head and face, throat, back and tail. White ventral area. Bright orange feathers, streaked with black, cover entire forehead, with longer outer plumes sweeping back over sides of head, some flaring behind eyes. Dark reddish-brown eyes. Large, sturdy, ridged orange-brown bill with noticeable bare pink skin at base, becoming triangular at gape. Pink feet with black soles.
Immature: Similar to adult, but slightly smaller. Dark grey throat. No plumes, or just a few yellow feathers randomly distributed on forehead. Less sturdy, dull, brownish bill with less noticeable bare skin at base.
Chick: Head and back dark grey-brown with off-white underparts.

SIZE AND WEIGHT
Body length approximately 71 cm; weight variable depending on gender and time of year, ranging between 3.1–6.6 kg.

VOICE
Contact call short, staccato and low-pitched. Courtship call loud, raucous and rhythmical braying, similar to Royal, but not as shrill as Southern rockhopper. Chicks emit simple cheeps which rise and fall when begging.

POPULATION AND DISTRIBUTION
Circumpolar, mainly on subantarctic islands. Global population estimated at 6.3 million breeding pairs, in approx. 55 breeding sites (c. 258 colonies), with the largest on Crozet (46°25'S), Kerguelen (49°21'S), South Georgia (54°18'S), and Heard and McDonald Islands (53°02'S).

A few pairs may breed among other species north to Falkland Islands and south to South Shetland

Islands. Rare sightings farther south in Antarctic Peninsula region, including failed breeding attempt near Anvers Island (64°33′S). In 2007, new high-latitude record from Avian Island (67°46′S) suggested possibility of range expansion due to climate-induced factors (Gorman et al, 2010).

Dispersive. Absent from colonies during pre-moult period and from April–May to October. Thirty adults from Crozet and Kerguelen in 2006/7 logged total distances travelled around 8000–10 000 km in six winter months, ranging roughly 1000–2000 km away from the colonies. (Thiebot at al, 2010).

Vagrant to New Zealand, Australia and Brazil.

BREEDING

Extremely social. Forms dense, noisy colonies, many numbering hundreds of thousands, established on level to steep rocky terrain usually with little or no vegetation (worn away by birds) or in tussock grass, depending on location. Much used routes facilitate access to nesting sites, which may be a few hundred metres above sea level. Breeding season September to March, variable depending on location. Males return to colonies from mid October to early November, with females following shortly afterwards. May commence breeding between ages 5–6 years, females usually about a year younger than males. High pair bond and nest site fidelity.
Courtship: Includes ecstatic display, swinging head from side to side, pumping flippers and braying loudly. Also includes mutual displays, bowing, quivering and mutual preening.
Nest: Small shallow depression or scrape in mud or gravel, lined with small rocks or grass.
Laying: Highly synchronous within colony, in November. 2-egg clutch, 4–5 days apart. First A-egg smaller than second B-egg.

Incubation: 33–37 days. Both birds remain at nest sharing incubation for first 12 days, then two more alternate shifts, female for about 12 days then male until hatching. A-egg is usually lost in early incubation. (See Morrison, p. 174.)
Chick rearing: Hatching normally from B-egg. Brooded/guarded by male for 20–25 days, female alone foraging and provisioning. Small crèches then form, with initial provisioning by female, then by both parents.
Fledging: 60–70 days, departing February-March.

FOOD

Varies depending on season and location, mainly krill, fish and some squid often at depths of 10–60 m (Heard Island) but capable of diving to well over 100 m.

PRINCIPAL THREATS

Predators: At sea, leopard seals and Antarctic fur seals. On land, giant petrels, skuas, gulls and sheathbills take unguarded chicks and eggs. Burgeoning fur seal population contributing to decline through competition for prey and breeding habitat.
Fisheries: Possible resource competition in some locations, especially krill.
Climate change: Declines linked to reduction in prey availability resulting from changing ocean conditions.
Disease: Although not always identified, epidemics can significantly affect local populations, e.g. two cases on subantarctic Marion Island: in 1993 an estimated 5000–10 000 penguins died from unknown disease at Bullard Beach; November 2004 approximately 2000 killed by avian cholera (*Pasteurella multicida*) outbreak at Kildalkey Bay. (Cooper et al, 2009).
Volcanic activity: Some very large colonies located on volcanically active islands at potential risk.

ABOVE Individuals nesting north of normal range sometimes pair up with the much smaller Southern rockhoppers, Kidney Island, Falklands.
LEFT A pair nesting amid Southern rockhoppers on New Island, Falkland Islands.
FAR LEFT Pair mutual preening on nest, Hannah Point, Livingston Island, South Shetlands.

BELOW LEFT With its sturdy stance, black face and flaming crown, the species is easy to recognise, South Georgia.
BELOW Flowing orange and black plumes stream from the centre forehead toward the sides and back of the head, South Shetland Islands.

Royal Penguin
Eudyptes schlegeli

Alternative or previous names: None known.
First described: Finsch, 1876.
Taxonomic source: Sibley and Monroe (1990).
Taxonomic note: Some authors consider *Eudyptes schlegeli* as subspecies of Macaroni penguin *Eudyptes chrysopholus*, but most treat these separately.
Origin of name: '*schlegeli*' after Hermann Schlegel, a German ornithologist and herpetologist, who was director of the Natural History Museum, Leiden.
Conservation status: Near threatened (IUCN 2018 — downlisted in 2015).
Status justification: Despite population being large and assumed stable, limited range and restricted breeding habitat increases vulnerability to human-induced or natural catastrophic events.

ABOVE AND BELOW
Although variable in some individuals, the white face, often tinged with yellow, is diagnostic for the species. All photos Macquarie Island, Australian Subantarctic.

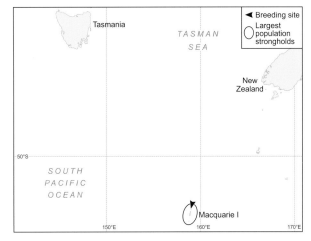

DESCRIPTION
Although closely related, adults unlikely to be confused with black-faced Macaroni penguin; immatures of both species have variable greyish throats, but usually paler in Royals. Before full crest development, juveniles may be confused with other crested species, especially at sea.

COLORATION
Adult: Black crown, back and tail. White ventral area. Face variable white to pale grey from just below crest, around eyes, to cheeks and throat. Bright yellow-orange feathers, streaked with black, cover entire forehead, with long outer plumes streaming back around sides of head, drooping behind eyes. Variable lemon yellow tinge around base of bill. Dark brown eyes. Large, reddish-brown bill, with noticeable bare pink skin outline becoming triangular at gape. Pink feet with black soles. Occasional dark morph with sooty grey throat and cheeks, resembling Macaroni. Though rare, some evidence of mixed species pairs and hybrid Royal/Southern rockhopper penguins.
Immature: Similar to adult but slightly smaller. Grey cheeks and throat. Dense pale yellow feathers on forehead but no long outer plumes. Smaller, dull, brownish bill with less noticeable bare skin at base. Adult plumage acquired about age 2 years.
Chick: Dark grey-brown head and back, with off-white underparts.

SIZE AND WEIGHT
Body length 65–75 cm; weight variable depending on gender and time of year, ranging between 3.0–8.1 kg.

VOICE
Similar to Macaroni penguin, extremely loud and vocal. Contact call short and low-pitched 'bark'. Courtship call includes harsh braying and 'trumpeting'. Chicks emit simple 'cheeps'.

POPULATION AND DISTRIBUTION
Breeding restricted to Australian subantarctic Macquarie Island, (54°37'S), with largest colony at Hurd Point and small numbers on adjacent Bishop and Clerk Islets (55°03'S).

Steady recovery subsequent to heavy human exploitation for oil in late 19th and early 20th centuries. Population considered stable; based on 1984–85 data, 850 000 distributed in 57 colonies. No recent census performed. In 2008, 457 breeding pairs were counted on Bishops Islet (Brothers and Leddingham, 2008).

Little known about winter dispersal, adults departing colonies after annual moult, mid to late April, and staying at sea until start of next breeding season. Breeders forage in deep offshore waters in polar frontal zone, travelling up to 160 km when chick rearing; round trips recorded to 1200 km during incubation.

Vagrant to New Zealand, Tasmania, Australia and South Georgia, suggesting wide oceanic movements.

BREEDING
Extremely social, breeds in dense and noisy colonies containing thousands of pairs, usually on rocky slopes and amid tussock-covered hills up to 200 m above sea level, some almost 2 km inland and accessed via creeks. Males return to colonies from mid September, females arriving early October. Breeding commences from about

age 5 years, but with lower success until about 10 years old. High long-term pair bond and nest site fidelity.

Courtship: Involves ecstatic and mutual displays, vertical head swinging, braying, bowing, quivering and mutual preening.

Nest: Shallow nest bowl, often lined with grass and small stones, or scraped in bare, level areas of sandy, pebbly or rocky ground.

Laying: Highly synchronous, from mid to end of October. Two white eggs, with approximate 4-day interval; A-egg usually displaced by female prior to laying of second, larger B-egg (see Morrison, p. 174).

Incubation: Between 32–37 days, shared by both parents in shifts of 12–14 days.

Chick rearing: Single chick raised. Brooded by male for first 3–4 weeks with female alone foraging and feeding chick. Crèches then form, with regular provisioning by both parents in 2–3 day shifts.

Fledging: Approximately 65 days. Chicks leave colony early to mid February.

FOOD
Diet varies between colonies on east and west coast of Macquarie Island, mainly consisting of krill and lantern-fish, with some squid and other small crustaceans, usually at depths of less than 60 m, but sometimes exceeding 100 m.

PRINCIPAL THREATS
Historical: In 2014, after long, arduous campaigns, Macquarie Is. was declared pest-free: previously feral cats, rats and mice took eggs and chicks; rabbits caused habitat loss through severe erosion triggering landslides and washouts.

Predators: Skuas take unguarded chicks and eggs.

Climate change: Recently identified as one of only two current threats to the species, through likely long-term effects on food supply (Trathan *et al.* 2015).

Disease: Identified as a potential threat, though considered unlikely given no historic precedent and good biosecurity measures (Trathan *et al.* 2015).

Disturbance/Tourism: Impact of tourism minimised by strict controls on tourist numbers and movements at only two permitted landing sites. ⬧

ABOVE Newly arrived bird preening, with crest still wet.
ABOVE LEFT Pair resting near shoreline.
LEFT (TOP TO BOTTOM) Commuters heading up muddy stream bed in Sandy Bay; moulting group about one month after the end of the nesting season; adults and chicks in extremely dense breeding colony.

BELOW Immature moulting into adult plumage.

Fiordland Penguin
Eudyptes pachyrhynchus

Alternative or previous names: Fiordland-crested penguin; Tawaki or Pokotiwha (Maori names).
First described: G.R. Gray, 1845.
Taxonomic source: Christidis and Boles (1994); Sibley and Monroe (1990); Turbott (1990).
Taxonomic note: Retained as separate species from Snares penguin *Eudyptes robustus* (Sibley and Monroe, 1990, 1994) contra Christidis and Boles (2008) who include latter as a subsp.
Origin of name: '*pachyrhynchus*' from Ancient Greek, 'pachy' meaning 'thick'; 'rhynchus', 'beak'.
Conservation status: Near threatened (IUCN 2020 — downlisted in 2020).
Status justification: Downlisted due to much higher than previous population estimates, though still suspected to be declining.

BELOW Like other crested penguins, this species is sure-footed on slippery boulders.
BELOW RIGHT A pair greets at the nest, showing diagnostic white lines on cheeks, Jackson Head, New Zealand.

DESCRIPTION
Most timid of seven in genus *Eudyptes* and rarest of New Zealand penguins. May be confused with other crested penguins, especially when wet or at sea, in particular Snares and Erect-crested due to size, stance and neighbouring distributions. Absence of bare pink skin around gape diagnostic.

COLORATION
Adult: Back and tail bluish-black; white ventral area. Head and throat black, with 3 to 6 contrasting thin white cheek lines in most individuals. Broad, yellow superciliary stripes start near base of bill, passing above eyes towards back of head, where slightly longer plumes form tidy drooping crest. Reddish-brown eyes. Sturdy, dark orange-brown bill without bare pink skin at base and gape as in similar species. Pink feet with black soles.
Immature: Similar to adult, but slightly smaller and slimmer. Shorter crest plumes do not droop. Chin and throat off-white or grey. Dark brownish bill.
Chick: Head, throat and back dark brown with off-white underparts.

SIZE AND WEIGHT
Body length 55–60 cm; weight variable, depending on time of year and gender, usually between 2.1–5.1 kg.

VOICE
Similar to Snares and Erect-crested penguins, but considerably less vocal habits. Calls peak when birds return to colony in early evening. Generally low-pitched but loud and harsh. Courtship calls include gradually lengthening throbs and low-pitched braying. Disagreements include hissing, growling and squealing. Chicks emit simple 'cheeps', rising and falling in pitch when begging.

POPULATION AND DISTRIBUTION
Endemic to New Zealand, winter breeder along southwest coast of South Island and outlying islands, from Bruce Bay (43°36'S) south to Fiordland and Solander Island (46°34'S), plus Stewart Island (47°00'S). Census and trends difficult due to small, fragmented colonies located in dense forest with difficult access; current population estimated higher than previous counts, at 12 500–50 000 mature individuals (Mattern & Wilson, 2019).

Dispersive. Absent from colonies during pre-moult and from end March to June, most birds travelling up to 3000 km southwest to feed along the subantarctic front.

Vagrant to all New Zealand's subantarctic islands,

and east coast of South Island; also Macquarie Island, Tasmania and occasionally southern Australia.

BREEDING

Winter breeding season July–November. Generally nests in small, loose colonies hidden from coastal view in temperate rainforest habitat, rocky shorelines or in caves, up to 800 m inland. Males return to breeding areas in June, with female following shortly afterwards. Breeding usually commences between ages 5–6 years. High pair bond and nest site fidelity, although often forced to relocate due to rain or flood damage.

Courtship: Involves 'vertical head swinging', mostly by lone males; mutual displays with braying, quivering, bowing and mutual preening.

Nest: Shallow scrape or hollow, preferably in soft muddy substrate, occasionally lined with sticks, stones, leaves, moss or fern fronds. Nests situated in dense vegetation or cavity among tree roots, also sheltered under rocks or in caves, often on steep slope.

Laying: 2 chalky white eggs, approximately 4 days apart, from end July to mid-August. First A-egg is smaller than second B-egg (see Morrison, p. 174).

Incubation: 30–36 days. Both adults remain at nest, sharing incubation for first 5–10 days, then alternate shifts of about 13 days, normally male first.

Chick rearing: Hatching from September. Usually only one chick raised, normally from B-egg. If smaller A-egg hatches (usually about a day later) chick often neglected and dies within first 10 days from starvation. Brooded by male for first 2–3 weeks, with female alone foraging and feeding chick. Small crèches may then form if other chicks nearby, with female initially continuing provisioning, then both parents.

Fledging: About 75 days, commencing mid-November. Develops adult plumage at about age 2 years.

FOOD

Little known, mainly squid, krill and small fish.

PRINCIPAL THREATS

Predators: At sea and on beaches, mainly Hooker's sea lions. On land, unleashed dogs, stoats and ferrets. Feral cats, possums and rodents appear not to have much impact.

Fisheries: Possible competition for prey species, especially squid. Potential mortality in trawls and occasionally set nets.

Climate change: Any changes in marine environment, e.g El Niño, affect prey species and have been shown to affect breeding success (Mattern & Ellenberg, 2016).

Disturbance/Tourism: Extremely susceptible to human disturbance which can adversely affect breeding birds, resulting in nest failure. Occasionally killed on roads.

Oil pollution: Potential for oil spills from coastal shipping. In 2021 New Zealand government prohibited further oil prospecting in NZ waters, reducing prior risks posed by expansion of this industry. ♠

ABOVE A commuter travels quietly through temperate rainforest to its hidden nesting colony, Westland. ABOVE LEFT Pair coming ashore at dusk, Murphy's Beach.

BELOW (THREE) Chicks are raised in the secrecy of dense, rain-drenched coastal forests along the west coast of New Zealand's South Island.

Snares Penguin
Eudyptes robustus

Alternative or previous names: Snares-crested penguin.
First described: Oliver, 1953.
Taxonomic source: Christidis and Boles (1994); SACC (2006); Sibley and Monroe (1990); Stotz et al (1996); Turbott (1990).
Taxonomic note: *Eudyptes robustus* and *Eudyptes pachyrhynchus* (Macaroni penguin) maintained as separate species contra Christidis and Boles (2008) who include *E. robustus* as a subspecies of *E. pachyrhynchus*.
Origin of name: 'robustus' meaning 'robust' in Latin, in this instance referring to sturdy bill.
Conservation status: Vulnerable (IUCN 2018 — uplisted in 1994).
Status justification: Thought to be relatively stable population. However, extremely small range, with breeding restricted to a single, small island group, makes species highly susceptible to stochastic (random probability) events, natural or anthropogenic, such as oil spills and fishing activities.

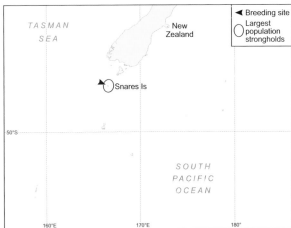

BELOW Pair along the edge of the *Olearia* forest. All photos taken on Snares Islands, New Zealand Subantarctic. BELOW RIGHT Mutual preening is an important part of pair bonding.

DESCRIPTION
Less aggressive than other Eudyptids. May sometimes be confused with others in genus, in particular Fiordland and Erect-crested, especially when wet or at sea. Often roosts on tree stumps and branches, sometimes up to 2 m above ground.

COLORATION
Adult: Head and upper throat black. Back and tail bluish-black; white ventral area. Narrow, yellow superciliary stripes start on either side of upper mandible, curving slightly over eyes, with longer crest feathers drooping and flaring outwards towards back of head. Bright reddish-brown eyes. Large, sturdy, ridged orange-brown bill, with conspicuous bare pink skin at base becoming triangular at gape. Pink feet with black soles.
Immature: Similar to adult, but slightly smaller and slimmer. Superciliary stripe initially paler, with shorter crest feathers. Throat off-white or grey.
Chick: Head and back dark grey with off-white underparts.

SIZE AND WEIGHT
Body length between 51–61 cm; weight variable depending on gender and time of year, ranging between 2.4–4.3 kg. Male bill substantially heavier than female.

VOICE
Similar to Fiordland penguin, though much noisier. Low-pitched, but loud, harsh and very persistent. Contact call a simple, mainly single 'bark'. Courtship calls include pulsing throbs which increase in length, turning to braying towards end of display. Disagreements includes hisses, squeals and groans. Chicks emit simple, repetitive 'cheeps' when begging.

POPULATION AND DISTRIBUTION
Found only in New Zealand's subantarctic region, where breeding restricted to islands in The Snares (48°01'S) group, with largest numbers on North East Island, Broughton and rocky Western Chain islets. A 2013 count reported 20 716 pairs on North East Island and 4433 pairs on Broughton Island (Hiscock & Chilvers, 2016), equating to 25 149 breeding pairs for the two main islands.

Considered pelagic and migratory. GPS studies show foraging trips up to 200 km east of The Snares during incubation period; shorter trips during chick rearing, within 80 km radius northwards (Mattern, 2013).

Absent from colonies during pre-moult/non-breeding periods, roughly March and May to August, with birds travelling up to 3500 km westwards into subtropical Indian Ocean, mainly north of 45°S (Thompson, 2016).

Vagrant to New Zealand, Chatham Islands, Falkland Islands, Macquarie Island, Tasmania and South Australia.

BREEDING

Extremely social, moults and breeds in noisy, dense colonies, ranging from just a few up to 1500 nests, up to 70 m above sea level. Sites include relatively flat or gently sloping muddy, peaty or bare rocky ground, often sheltered by bushes, tussock grass or under forest canopy of tree daisies *Olearia lyalli* and *Brachyglottis stewartiae*. Intense activity during breeding season tends to create quagmire conditions so, where possible, colonies slowly relocate to nearby intact areas. Breeding assumed to commence at about 6 years old. On North East Island, males usually return to colonies in early September, female following about one week later. On Western Chain islets timing of breeding delayed by up to 6 weeks. High nest site and pair bond fidelity season to season.

Courtship: Involves vertical head swinging, braying with 'forward' and 'vertical' bowing, quivering and mutual preening.

Nests: Shallow, often soggy nest bowl, lined with twigs, mud, stones and bones.

Laying: Two-egg clutch, in late September–early October. The first, A-egg, is smaller than second, B-egg, laid about 4 days later.

Incubation: Shared by both parents in three shifts for 31–37 days, usually beginning after B-egg is laid. Both remain at nest for first 10–14 days sharing incubation duties, then switching to shifts of approximately 12 days, beginning with female.

Chick rearing: Hatching early November onwards, single chick raised, usually from larger B-egg. If both eggs hatch, smaller chick is neglected and starves (see Morrison, p. 174). Brooded by male for first 2–3 weeks, with female alone foraging and feeding chick. Small crèches then form, with both parents provisioning, although female more regularly.

Fledging: About 75 days, departing mid-January to mid-February from North East Island and up to March from Western Chain islets.

FOOD

Mainly krill, squid and small fish usually at depths of 26 m or less but sometimes up to 120 m.

PRINCIPAL THREATS

Predators: At sea, Hooker's sea lions; more rarely, leopard seals. On land, skuas may take unattended eggs/chicks. No introduced predators, though potential for accidental introduction remains.

Habitat: Well protected and unspoilt. New Zealand Government prohibits landings, except for infrequent scientific study, keeping human disturbance minimal.

Fisheries: Large-scale squid fishery near breeding islands may cause competition for prey. Also, set net bycacth is known source of mortality.

Climate change: Potential threat from changing ocean conditions and resultant reduced prey availability.

Oil pollution: Potential for oil spills from shipping.

ABOVE Adults heading down granite boulders to the shore.
ABOVE LEFT Group arriving on shore.
BELOW Drinking from a rain puddle.
BOTTOM LEFT Group moulting in the sheltering *Olearia* forest.
BOTTOM RIGHT Commuters making their way through slippery bull kelp.

Erect-crested Penguin
Eudyptes sclateri

Alternative or previous names: Big-crested penguin, Sclater's penguin. Formerly known as *Eudyptes atratus*.
First described: Buller, 1888.
Taxonomic source: Christidis and Boles (1994, 2008); SACC (2006); Sibley and Monroe (1990); Stotz et al (1996); Turbott (1990).
Taxonomic note: Formerly regarded as a race of Fiordland penguin (*Eudyptes pachyrhynchus*).
Origin of name: 'sclateri' after the British zoologist Philip Lutley Sclater.
Conservation status: Endangered (IUCN 2020 — uplisted in 2000).
Status justification: Population estimates indicate significant declines of at least 50% from 1978 to 1998, with primary breeding range restricted to just two island groups. Probably still decreasing.

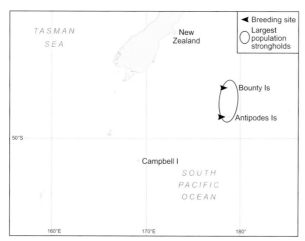

DESCRIPTION
One of the least studied penguins. Can be confused with other crested penguins, in particular Snares and Fiordland, especially when wet or at sea, but has a distinctive profile with domed crown, longer, thinner bill with less noticeable pink skin at base, and thick-set throat. Unlike all others, the bristly crest can be raised and lowered voluntarily when dry.

COLORATION
Adult: Back and tail bluish-black to jet-black; white ventral area. Head and upper throat jet black. Broad, yellow-gold superciliary stripes start near gape, passing over eyes to form long, parallel, brush-like crests. Flippers have largest amount of black on underside compared to other crested species. Dark reddish-brown eyes. Long, slender, orange-brown bill, with bare pink skin at base and gape. Pink feet with black soles.
Immature: Similar to adult, but slightly smaller and slimmer. Superciliary stripe paler, with shorter, non-erectile crest. Throat off-white or grey. Brownish-black bill with occasional pale tip; bare skin at base less noticeable.
Chick: Head and back greyish-brown, with white underparts.

SIZE AND WEIGHT
Body length 60–67 cm. Weight variable, depending on time of year and gender, usually between 3.0–7 kg.

VOICE
Similar to Snares and Fiordland penguins, but more vocal, very loud braying screeches, with low-pitched throbbing. Courtship calls accompanied by specific visual displays, with pulsing 'trumpet' call decreasing in pitch. Disagreements involve simple, non-pulsing yell, sometimes accompanied by hiss or groan. Chicks emit simple repetitive 'cheeps', which rise and fall in pitch when begging.

RIGHT Bare trampled ground among dense tussock grass marks the overcrowded Orde Lees colony along the western slope of Antipodes Island. BELOW Immature with grey throat and shorter crest, Campbell Island.

POPULATION AND DISTRIBUTION

Found only in New Zealand's subantarctic region, where breeding restricted to Antipodes (49°41'S) and Bounty (47°45'S) islands. Stragglers at Auckland (50°42'S) and Campbell (52°32'S) islands. Most recent survey in 2011 estimated 26 000 breeding pairs on the Bounty Islands (Miskelly, 2013) and found 34 226 nests on Antipodes Islands (Hiscock & Chilvers, 2014), which extrapolates to c.150 000 mature individuals for the species (80% engaged in breeding).

Dispersive and thought to be pelagic; absent from colonies during pre-moult period, and afterwards mid April to September. Very little known about migratory pattern.

Vagrant to North and South Islands of New Zealand, Tasmania, South Australia, Macquarie, Kerguelen and Falkland Islands.

BREEDING

Extremely social, moults and breeds in large, noisy colonies of many thousands. Nests amid Salvin's albatross, Fulmar prions and New Zealand fur seals on Bounty Islands, and shares space with small numbers of Southern rockhopper penguins on Antipodes.
Dense colonies established on rugged open ground from shoreline to 75 m above sea level, often in rough terrain with precipitous access. Breeding assumed to commence at about 4 years old. Males return to colonies in early September, females about 2 weeks later. Presumed monogamous, with long-lasting pair bonds.
Courtship: Involves ecstatic greetings, vertical head swinging, 'forward' and 'vertical' trumpeting, bowing, quivering and mutual preening.
Nests: Shallow nest bowl rimmed with small stones. Grass sometimes used as lining if available; built on flat or slightly sloping ground, or between boulders.

Laying: Two-egg clutch, early October with peak laying mid-month. First, A-egg (chalky pale blue or greenish) is much smaller than second, B-egg (off-white), laid about 4 days later.
Incubation: Shared by both parents for about 35 days, beginning after B-egg is laid. A-egg is usually lost within first 3–4 days of incubation – presumed intentionally ejected or accidentally dislodged (see Morrison, p.174).
Chick rearing: Single chick raised, hatching about mid-November, usually from larger B-egg. Brooded by male for first 2–3 weeks, with female alone foraging and feeding chick. Small crèches then form, with both parents provisioning, although female more regularly. 'Food chases' sometimes occur when parent seemingly leads chick away from crèche to be fed.
Fledging: About 70 days, departing end of January to mid-February.

FOOD

Assumed to consist mainly of krill, other crustaceans and squid.

PRINCIPAL THREATS

Historically hunted for skins.
Predators: Sporadic killing by fur seals has been observed (Antipodes Island). Skuas take unattended eggs and chicks. Million Dollar Mouse campaign eradicated introduced mice from Antipodes in 2018.
Habitat: Nesting grounds well protected and unspoilt, as New Zealand Government prohibits landings, except for infrequent scientific study, human disturbance minimal.
Climate change: Factors affecting marine environment and prey species, such as warming sea temperatures thought to be instrumental in species decline.

ABOVE Nesting amid Salvin's albatrosses, space on Bounty Islands, the species stronghold, is at a premium.
ABOVE LEFT A secluded nest site on the periphery of the colony, Antipodes Island.

LEFT (BOTH) This is the only species able to erect or flatten its crest at will.

Yellow-eyed Penguin
Megadyptes antipodes

Alternative or previous names: Hoiho (Maori name meaning 'noise shouter'); ancient Maori name 'Tavora'.
First described: *Catarrhactes antipodes,* Hombron and Jacquinot, 1841.
Taxonomic source: Sibley and Monroe (1990); Turbott (1990).
Taxonomic note: Monotypic. No subspecies but mainland and island populations genetically distinct.
Origin of name: *'antipodes'*, referring to New Zealand being geographically opposite Europe.
Conservation status: Endangered (IUCN 2020 — uplisted in 2000).
Status justification: Very small breeding range with severe habitat degradation in many areas. Extreme population fluctuations, plus severe declines in mainland regions.

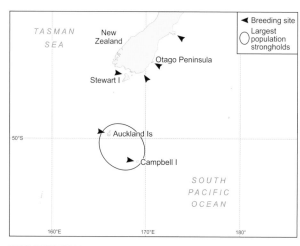

VOICE
Very little vocalization, except at nest. Contact call is a two-syllable, high-pitched sound. Extremely loud courtship and greeting calls are a repetitive and high-pitched staccato braying. Disagreements or threats induce yelling, or series of harsh, gutteral 'chuckles'. Chicks emit simple, repetitive 'cheeps'.

POPULATION AND DISTRIBUTION
Found only in New Zealand, along parts of southeast coast of South Island (mostly below 43°44'S), Stewart Island (47°00'S), Codfish Island (46°46'S) and other outlying islands in Foveaux Strait, plus largest populations on subantarctic Campbell (52°32'S) and Auckland (50°40'S) islands. Various counts over several years yield an estimated 2684–3064 mature individuals.

Highly sedentary, adults foraging along coast in vicinity of breeding areas, to where they return at night. During breeding season, most foraging trips within 20km of nest site, with longer distances up to 60 km during incubation and post-guard periods. Some movements northward up to 500 km (as far as Cook Strait), mainly juveniles.

DESCRIPTION
The only extant species in this genus and one of the rarest, most endangered penguins. It is the least social and quite timid. Unique appearance so no risk of confusion with other penguins.

Closely related to extinct Waitaha penguin (*Megadyptes waitaha),* smaller member of genus whose existence was only recently discovered following DNA analysis of 1000-year-old bones (see Boessenkool, p. 164).

BREEDING
One of longest penguin breeding cycles, mid-August through March, with some courtship activity starting earlier. Nests located on coastal slopes and sand dunes, open forest and pasture, amongst tussock grass, shrubs, gorse and flax. Favoured sites in coastal podocarp/hardwood forests, but little such habitat remains.

COLORATION
Adult: Bluish-black, slightly grey nape, back and tail with white ventral area. Head pale yellow, streaked with black from bill to top of crown, less so around bill base and under eyes. Well-defined, lemon-coloured band runs around back of crown and diagonally forward to encircle each eye. Slightly erectile yellow and black crest feathers. Lower cheeks, throat and side of neck yellowish-brown. Bill mostly orange-red upper mandible with some pink in middle and at base; lower mandible pink with orange-red tip. Pale yellow eye with pink rim. Pink feet with black soles.
Immature: Similar to adult but head all yellow-grey, without band, and shorter crest feathers. White throat and ventral area. Eyes lemon-grey.
Chick: Entirely dark brown.

SIZE AND WEIGHT
Body length between 56–78 cm; weight variable depending on gender and time of year, ranging between 3.6–8.9 kg.

Loosely colonial with large territories, preferring to nest in privacy, well away and out of site of congeners. Commence breeding between 2–5 years old, females earlier than males. Strong attachment to nesting area (but not nest site), with high pair fidelity during lifetime.

Courtship: Includes ecstatic display, with sky-pointing and 'trumpeting' most often by lone male at nest site, also bill quivering, braying duets and mutual preening.

Nest: Shallow nest bowl made with twigs, leaves, grass or any other available vegetation, built by both birds and usually sited against solid backdrop such as tree base, rock, flax plant. Also readily uses man-made nesting shelters placed in key nesting areas by conservation groups.

Laying: Normally 2 bluish-green eggs (change to white after about 24 hours) between mid-September and mid-October. Laying interval 3–5 days.

Incubation: 38–54 days, shared by both partners. May commence immediately upon completion of clutch, or up to 10 days delay before full incubation.

Chick rearing: Usually both hatch on same day, in early November. Duties shared by both adults. Chicks do not form crèche.

Fledging: 106–108 days, departing in late February to mid March. Moult to adult plumage between 14–16 months.

FOOD

Forages inshore and, unlike other penguins, almost exclusively along seafloor. Diet varies depending on season and location but mostly consists of red cod, opal fish, sprat, silversides, blue cod and ahuru, with some squid and krill, usually at depths of less than 40 m but capable of diving in excess of 150 m.

PRINCIPAL THREATS

Predators: At sea, Hooker's sea lions, barracouta and sharks. In some areas, predation of chicks and eggs by feral cats, stoats, ferrets and feral pigs cause serious impact. Unleashed dogs known to attack/harass adults.

Habitat: In the South Island, forest nesting habitat devastated by logging, occasional fires, land clearance for agriculture, development and recreation, plus grazing and trampling of nest sites by livestock.

Disease: Potential for drastic effects on overall population. For example, in 2004 an outbreak of diphtheric stomatitis (avian diphtheria) resulted in 60% chick mortality between Oamaru and Stewart Island; 2007, a blood parasite on Stewart Island appeared to cause 100% chick mortality.

Fisheries: Accidental bycatch in commercial gillnetting results in substantial mortality. Trawling, oyster fisheries and bottom longlining may cause benthic habitat degradation affecting quality and availability of prey (Ellenberg & Mattern, 2012), and seafloor foraging behaviour (Mattern *et al.*, 2013).

Climate change: Population decline in large parts of range, where chick starvation and disease noted. Changing sea surface temperatures appear to influence productivity and corresponding survival rates.

Disturbance/Tourism: Adults shy and easily frightened by human presence on beaches, which may inhibit access to nests, causing reduced feeding of chicks, and abandonment. Access monitored in popular areas.

ABOVE (BOTH) Solitary nesters in dense subantarctic vegetation, Campbell (left) and Enderby (right) Islands.

ABOVE FAR LEFT Commuter returning to sea, passing Hooker's sea lion breeding colony, Enderby Island.

BELOW Nesting inside artificial shelter provided at Penguin Place, private penguin reserve on Otago Peninsula.

BELOW CENTRE Pair near shore in Pohatu Marine Reserve, Flea Bay, Banks Peninsula.

BELOW LEFT Immature prospecting potential nest site, Campbell Island.

Little Blue Penguin
Eudyptula minor

Alternative or previous names: Little penguin, Blue penguin; Korora (Maori); Fairy penguin (Australia).
First described: *Aptenodytes minor,* J.R. Forster, 1781; *Eudyptula minor albosignata* Finsch, 1874.
Taxonomic source: Christidis and Boles (1994, 2008); SACC (2006); Sibley and Monroe (1990); Turbott (1990).
Taxonomic note: Formerly two species recognised, Little penguin (*Eudyptula minor*) and Fairy penguin (*Eudyptula undina*), now accepted as single *E. minor*. Currently 6 subspecies as follows, Australia: *E. m. novaehollandiae*, southern coasts and islands (west to east, Fremantle to Port Stephens, including Tasmania); New Zealand: *E. m. minor*, southern half of South Island (Karamea to Oamaru); *E. m. iredalei*, eastern North Island (Kawhia to East Cape); *E. m. variabilis*, northwestern North Island (Cape Egmont to Hawkes Bay); *E. m. chathamenis*, Chatham Islands; *E. m. albosignata,* (White-flippered penguin) regarded by some as distinct species restricted to Banks Peninsula, eastern South Island.
Origin of name: *'minor'* from Latin, 'smallest'; 'blue' due to unique plumage colour (see Shawkey, p. 162).
Conservation status: Least Concern (IUCN 2020 — since 1988).
Status justification: Maintains large range. Some local population declines and extinctions, but widespread population trend is generally regarded as stable and above the 'Vulnerable' status threshold.

ABOVE (BOTH) Differences in size and plumage colour are obvious between the widespread race (top), seen here in Tasmania, and the White-flippered subspecies (bottom) found only around Banks Peninsula on the eastern coast of New Zealand's South Island, Pohatu Penguins, Flea Bay.

BELOW Plumage coloration appears turquoise when wet, Phillip Island, Australia.

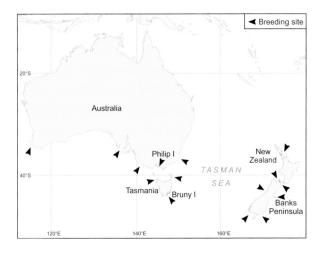

Immature: Similar to adult but dorsal feathers more vivid; slimmer, shorter bill.
Chick: Greyish-brown head and back with off-white underparts.

SIZE AND WEIGHT
Only slight variation between genders; heavier bill in male. Body length 40–45 cm; weight variable depending on gender and time of year, ranging between 1– 1.7 kg.

VOICE
Most vocal at night. Contact call short and high pitched. Advertising call consists of rhythmical braying produced while exhaling (gutteral, throbbing sound) and inhaling (short, high-pitched squeal), usually by lone male. Variations used in courtship and greeting duet. Squabbles involve growls, brays, barks, yells and hissing. Chicks emit high-pitched 'peeps'.

POPULATION AND DISTRIBUTION
Coastal mainland and offshore islands: southern Australia and Tasmania; entire New Zealand coastline Northland to Stewart Island (47°00'S), except Fiordland and South Westland, plus Chatham Islands (44°02'S). Population counts difficult, but approximately 469 760 breeding adults estimated overall. White-flippered penguin restricted to Banks Peninsula (43°45'S) and nearby Motunau Island (43°03'S), New Zealand.

Sedentary year round, forages up to 25 km offshore mostly within 70 km of colony. Often seen in bays, harbours and estuaries. Some juvenile dispersal over 600 km northwards.

Vagrant to southeastern Australia, Lord Howe and Snares Islands.

BREEDING
Sociable, in small, loose colonies, rarely solitary pairs. Prefers well-vegetated sandy or rocky coastal slopes, also lightless caves, up to 300 m above sea level and 1.5 km inland. Start of breeding season varies by

DESCRIPTION
Smallest penguin, nocturnal on land, returning to burrows in small groups at or after nightfall, departing to forage just before dawn. In undisturbed areas, may rest on shore in daylight. Subspecies easily confused but very little range overlap.

COLORATION
Adult: Upper half of head, back and tail variably metallic midnight-blue, blue-green or duller greyish-blue depending on race, light and if wet/dry. Dorsal feathers have off-white base and black shafts, noticeable at close quarters. Varying pale greyish-blue ear coverts and chin fading to silvery-white throat and ventral area. Flippers dark blue-black uppers, with variable white trim; pale below. Eyes vary from pale to grey-blue. Dark, almost black, short bill with paler underside. Pale pink feet with dark soles. White-flippered penguin slightly larger and paler, with mostly white underside, dusty grey-blue back and broad white flipper edges, white joining at elbow on some males.

location, generally July–December. Commence breeding from age 2–3 years. High pair bond and nest site fidelity.

Courtship: Ecstatic display initially by lone male, stretching tall and sky-pointing with braying, usually by prospective burrow. Similar displays in bonding pair, plus bowing, walking around burrow together and mutual preening. Cave breeders may perform in small, central group.

Nest: Cavity lined with grass, leaves, twigs, bones, rocks, seaweed or other vegetation, either in burrow dug in soft soil or sandy substrate, or other hollows, e.g. bird or rabbit hole, under rocks or vegetation, in crevices, pipes, even man-made structures or buildings; also open nests within unlit caves. Readily adopt artificial nest boxes.

Laying: 2-egg clutch, 2–3 days apart, mostly July–December, peaking August–November, but highly variable. Early breeders may double clutch.

Incubation: 33–39 days, shared by both parents.

Chick rearing: Hatch synchronously, usually within a day. Two chicks normally raised. Brooding and guarding by both parents for about 3 weeks, then left unattended in burrow during daytime foraging. Small crèches form only in cave nesting colonies.

Fledging: 7–9 weeks.

FOOD

Diet varies depending on location. Small fish including anchovy, sprat, pilchards, juvenile red cod, blue warehou, also squid and small crustaceans, usually at less than 20 m, with deepest recorded dive 72 m.

PRINCIPAL THREATS

Predators: At sea, mainly sharks and sea lions, New Zealand fur seals. On land, dogs are primary killers, also feral cats, stoats and ferrets. On NZ West Coast weka take chicks. Also foxes in Australia. Quolls and Tasmanian devils in Tasmania. In 2012/13 aggressive facial tumour disease was decimating the endangered Tasmanian devil population, so 28 healthy individuals

were released as insurance on Maria Island, a National Park and Marine reserve off the east coast of Tasmania. In a conservation paradox, the subpopulation thrived but at the catastrophic expense of the island's birdlife, including the loss of 3000 breeding pairs of Little blue penguin. As of an early 2020 Parks and Wildlife Service survey, all penguin colonies were empty.

Fisheries: Competition for prey species, plus pollution from aquaculture. Accidental by-catch in recreational set-nets cause of major localised mortalities.

Habitat: Increasing encroachment of nesting areas from vegetation clearance and coastal development (housing, agriculture, roading, port facilities, etc.).

Climate change: Changing ocean conditions may exacerbate reduction in prey availability, as in 1997 collapse of southern Australia's sardine stocks (see Chiaradia p. 178). Increasing land temperatures in breeding season may cause mortality in chicks/adults due to hyperthermia, especially those nesting under vegetation rather than in burrows.

Disturbance/Tourism: Human activity on beaches near nesting areas disrupts and delays access for feeding of chicks, mate relief, etc. Concurrent mortality from traffic, plus domestic cat and dog harassment and predation may result in abandonment of site.

Oil pollution: In 2021 New Zealand government prohibited any further oil prospecting in NZ waters, reducing prior risks posed by expansion of this industry. Accidental oil spills caused by grouding of commercial shipping, often due to navigational errors. Examples include: 1995 grounding of *Iron Baron* on Hebe Reef, Tasmania, resulted in estimated 10 000–20 000 penguins dead at sea, with a further 1894 rescued and de-oiled; 2011 grounding of MV *Rena* on Astrolabe Reef, New Zealand, with a tally of 383 live and 89 dead oiled adults collected from the region's total 200–300 breeding pairs.

ABOVE Chicks beg frantically to be fed by a freshly returned parent just outside of their nesting burrow, Bruny Island, Tasmania.

ABOVE LEFT Under the cover of darkness, adults in large numbers wend their way up the slope of Phillip Island, near Melbourne, Australia, where a tourism-funded conservation enterprise has spurred a population rebound.

LEFT Growing bolder as it begins to moult out of its baby down, a well-fed chick stretches outside its burrow, Bruny Island, Tasmania.

BELOW When the container ship *Rena* ran aground in the Bay of Plenty, over half of the local breeding population was directly affected.
PHOTO COURTESY OF MARITIME NEW ZEALAND

BELOW LEFT Although generally nocturnal in habits, in places where they don't feel threatened, landings begin during twilight hours, Phillip Island, Australia.

Here is a selection of some of the best places we have found where penguins can see people, like this Chinstrap on Deception Island, Antarctica.

WHERE TO SEE WILD PENGUINS

Australia A thriving colony of **Little blue penguins** at Phillip Island Nature Parks, near Melbourne, attracts visitors from around the world, with tourist funds turned back into protection and research. This penguin can also be seen coming ashore at night on many of Australia's southern beaches, for example The Neck Nature Reserve on Bruny Island near Hobart, Tasmania.
Season and contacts: September–January
www.penguins.org.au/attractions/penguin-parade
www.brunyisland.net/Neck/neck.html

New Zealand The South Island offers opportunities to see three very special species. Near Christchurch, you can experience a private guided walk to see the unusual **White-flippered penguins** (a subspecies of the Little blue) at Pohatu Penguins, on the tip of Banks Peninsula. Oamaru offers dusk viewing of homecoming **Little blue penguins**, and, further south, from Otago Peninsula to the Catlins, the rare **Yellow-eyed penguins** can be spotted coming ashore at various beaches in the early morning or late afternoon. Great care must be taken to remain out of sight at considerable distance, as they are extremely shy. Closer views can be had at Penguin Place near Dunedin, a working farm dedicated to penguin protection. The elusive **Fiordland crested penguin** nests in small colonies hidden in thick coastal forest in Westland. Look for telltale tracks at beaches all the way to the end of the road at Jackson Head and beyond. The surest sightings are by taking a guided outing from the Wilderness Lodge at Lake Moeraki.
Season and contacts: (White-flippered) September–January, www.pohatu.co.nz; (Little blues) October–January, www.penguins.co.nz; (Yellow-eyed) November–February, www.penguinplace.co.nz; www.yellow-eyedpenguin.org.nz; (Fiordland crested) August–November, www.wildernesslodge.co.nz/wildernesslodge/lake-moeraki

Patagonia The best-known place to see **Magellanic penguins** is Punta Tombo (near Trelew) in Argentina, but other sites include Cabo Dos Bahías (near Camarones) farther south, as well as tours from Ushuaia to Isla Martillo in the Beagle Channel. In southern Chile, a drive from Punta Arenas to Seno Otway offers close-up encounters along well laid-out walkways or take a boat tour from Punta Arenas to visit the large colonies on Isla Magdalena in the Strait of Magellan.
Season and contacts: November–February
www.patagonia-argentina.com/en/punta-tombo

Falkland Islands This is the do-it-yourself penguin capital of the world, with four species — **Southern rockhopper, Magellanic, Gentoo, King** — guaranteed simply by making private arrangements and flying to some of the outlying islands, or taking an overland trip from Stanley by Land Rover. The largest and most accessible King penguin colony is at Volunteer Beach.

Elizabeth Bay, Galapagos Islands.

Neko Harbour, Antarctic Peninsula.

Cabo Dos Bahías, Argentina.

Camp with a view, Falklands.

Many Antarctic tour ships also include the Falklands to add variety and species, but a private stay is by far the most rewarding.
Season and contacts: December–February
www.falklandislands.com

Antarctica and South Georgia With a little luck, even a relatively short trip to the Antarctic Peninsula should offer good encounters with the three long-tailed penguins, **Adélie, Gentoo** and **Chinstrap**, nesting throughout the South Shetland Islands and even on the continental shores in places. Adélies tend to be furthest south or where sea-ice is most prevalent, such as Paulet Island in the Antarctic Sound or Petermann Island south of Lemaire Channel, whereas Gentoos are in the most sheltered bays, and Chinstraps centred between. South Georgia Island is unrivalled for its gigantic colonies of **King penguins**, with smaller numbers of Gentoos and Chinstraps. There are also large colonies of **Macaroni penguins**, but their exposed locations make access quite difficult.
Season and contacts: November–January. Numerous cruise ships and expedition yacht charters based in different countries offer options, therefore none in particular are proposed here. Smaller vessels (<100 passengers) tend to offer better shore opportunities.

Atacama Desert Coast, Peru-Chile The only places to see **Humboldt penguins** are on small desert islands not frequently visited. A few hints are Paracas Islands off the Paracas Peninsula in Peru; Choros, Chañaral and Pan de Azucar off north-central Chile; and Algorrobo Island (connected by locked causeway) near Valparaiso, Chile. None are easy to reach.
Season and contacts: April–December; http://mesh.biology.washington.edu/penguinProject/Humboldt

Galapagos Islands The Galapagos penguin is both one of the rarest penguins as well as the most often seen, by taking a boat cruise around these islands. Since this penguin is only ever seen in small numbers, to maximise chances of a good encounter, the cruise itinerary must include at least some of these stops: Bartolomé Island, Tagus Cove, Elizabeth Bay, Puerto Villamil's Las Tintoreras.
Season and contacts: May–December
www.Galapagostravel.com

East Antarctica and the Ross Sea This being 'deep' Antarctica, only two penguin species can be expected

here, **Adélie** and **Emperor**, the latter of which nests on the sea-ice in winter and is therefore unpredictable during the summer cruising season. The largest Adélie colony in the world is at Cape Adare, a sight to behold. But seeing Emperors with chicks requires an icebreaker and/or helicopter to reach the colonies before they disband.
Season and contacts: December–February
www.heritage-expeditions.com,
www.auroraexpeditions.com.au

Subantarctic Islands of Australia and New Zealand   Several unique crested penguins are found exclusively in this sector of the world: **Royals** on Macquarie Island; **Snares** on The Snares; and **Erect-crested** on Bounties and Antipodes Islands. Few passenger vessels ply these stormy seas, and those that do are only permitted to land on Macquarie. The other islands are weather dependent and are, at best, only visited by Zodiac tender, without actual landings. Still, when the gamble pays off, it's a superlative experience. Enderby in the Auckland Islands is also a great place to see **Yellow-eyed penguins.**
Season and contacts: December–February
www.heritage-expeditions.com; www.tiama.com

South Africa   There aren't many places in the world where penguins can be met simply by driving a short way out of the suburbs. Cape town is such a place. Simply follow the signs west to Simonstown and park at the penguin-viewing visitor centre at Boulders Beach, where boardwalks will lead you to the **African penguins.**
Season and contacts: June–September
www.sanccob.co.za

Tristan da Cunha   One of the least known and hardest to see of all penguins is the **Northern rockhopper.** Only very rarely does a passenger vessel visit some of the islands where they live in the southern Indian Ocean or mid-south Atlantic. Another way is a private visit to Tristan da Cunha Island, renown as the most isolated community in the world, by arranging passage on a fishing vessel from South Africa. Patience, flexibility and a sense of adventure are a must.
Season and contacts: October–December
www.tristandc.com

Elephant Island, South Shetlands.

Bruny Island, Tasmania.

NOTES ON REFERENCES AND FURTHER READING

While compiling this book, we have endeavoured to incorporate the latest and most accurate information available on penguin biology and life histories. Our quest for the most intriguing facts and lore has taken us on fascinating, though often circuitous, routes to many diverse sources, from antique tomes written in Old French or Latin to advanced research papers on quantum physics and fluid dynamics. Innumerable books, scientific journals and authoritative online sources have been consulted, and our computer browser histories show well over 100 000 web pages have been viewed in the process. With our readings touching on most every discipline of the sciences, we have tried, wherever possible, to relay interesting information using words understandable to non-specialist audiences. In addition, we have drawn extensively on our own observations, experiences and images gathered over many years of expeditions to penguin habitats, melding and cross-referencing this personal information with solid research data from peer-reviewed articles and scholarly publications. Unfortunately, space constraints have prohibited inclusion of an exhaustive bibliography and citations list, though some of our expert contributors in Part II were able to cite some of their own data sources; plus some included in Part III. However, while much of the text has been either reviewed or contributed by researchers and professionals in their fields, any errors and omissions, of course, remain entirely our own.

Following is a small selection of our reference titles, chosen for their interest and breadth of information, and that we feel offer a good starting point for individual research and further reading.

del Hoyo, Josep et al (Eds). *Handbook of the Birds of the World.* Vol 1. Lynx Edicions, 1992.
Lindsey, Terence & Morris, Rod. *Field Guide to New Zealand Wildlife.* Auckland: Harper Collins Publishers, 2000.
Reader's Digest Antarctica. Australia, 1985. Used for historical references.
Reilly, Pauline. *Penguins of the World.* Australia: Oxford University Press, 1994.
Ryan, Peter (Editor). *Field Guide to the Animals and Plants of Tristan da Cunha and Gough Island.* Newbury: Pisces Publications, 2007.
Shirihai, Hadoram. *A Complete Guide to Antarctic Wildlife* (2nd ed). Illustrated by Brett Jarrett. London: A&C Black, 2007.
Taylor, Rowley. *Straight Through from London*, Heritage Expeditions 2006. History of Bounty and Antipodes Islands.
Williams, Tony D. *The Penguins.* Illustrated by J.N. Davies & John Busby. New York: Oxford University Press, 1995.

Salisbury Plain, South Georgia.

Additionally, many websites offer detailed information:

www.arkive.org Images and general information on all species
www.birdlife.org/datazone Individual species factsheets
www.bluepenguin.org.nz The Blue Penguin Trust
www.falklandsconservation.com/wildlife/penguins Falklands Conservation
www.globalpenguinsociety.org Global Penguin Society
www.iucn.org/redlist IUCN Red List of threatened species
www.mesh.biology.washington.edu/penguinProject Penguin Project at Washington University
www.penguin.net.nz New Zealand Penguins
www.penguins.cl International Penguin Conservation Work Group
www.photovolcanica.com Images and general information on all species
www.yellow-eyedpenguin.org.nz Yellow-eyed Penguin Trust

Mahalia and crew, The Snares, New Zealand Subantarctic.

Amanda Bay, East Antarctica.

ACKNOWLEDGEMENTS

An incredible number of people have helped and encouraged us in the production of this book, all of whom we'd like to acknowledge, even if space may prevent individual mentions.

We firstly wish to thank those with whom we shared time in the field: researchers and conservation workers from the Galapagos to the Antarctic, Tristan da Cunha to Chile, Australia, New Zealand and South Africa; and the wonderful friends and fellow adventurers who helped us crew our little yacht *Mahalia* to mythical subantarctic islands: Grant, Carl, Jacinda, Ali, Alan, Julian, Andy, Chris, Sarah and especially Pete, with whom we never stopped laughing during those first four months of wind and wet. A special thank you to Meri Leask of Bluff Fisherman's Radio for being our faithful and tireless communications link during all our sailing expeditions. Also, Henk Haazen, of *Tiama*, outstanding yachtsman and fount of personal inspiration and friendship on the high seas; and Rodney Russ of Heritage Expeditions for sharing his vast knowledge of the subantarctic islands.

Many people helped us in various ways to spend time with penguins. In New Zealand, Paul Sagar, Pete McClelland, Greg Lind, Neville Peat, Kerry-Jane Wilson, Inger Perkins, the McGrouther family of Penguin Place on the Otago Peninsula, and many others facilitated encounters with our home country's rare and unusual penguins. Several remarkable people introduced us to the secretive night realm of the Little blue penguins: in New Zealand, Francis and Shireen Helps of Pohatu Penguins hosted us at their private White-flippered penguin haven on Banks Peninsula, while Reuben Lane led us to the cave-nesting Little blues on the rugged West Coast; in southern Australia, we are particularly grateful to long-time friends David Parer and Liz Parer-Cook, and the brilliant team at Phillip Island; likewise in Tasmania, thanks to dear friend Louise Crossley and her co-conservationists on Bruny Island.

Deserving of our very special appreciation is the Australian Antarctic Division, which embraced our project and invited Tui on a six-week icebreaker expedition deep into the ice on the 'Far Side' of Antarctica to photograph Emperor penguins in the spring, when few people have access to their colonies. In the process she witnessed first-hand the remarkable array of cutting-edge science being conducted in diverse and fascinating fields by this long-running programme. Patti Lucas, Sharon Labudda, Leanne Millhouse, Alison Dean, Mike Grimmer and all the expeditioners and crew aboard the R/V *Aurora Australis* are remembered for their friendship and cheerful help.

We thank our many friends in the Falkland Islands, particularly Phil and Stella Middleton in Stanley and the Delignières family of Dunbar Farm, for opening up their homes and their hearts to us. Also Ian and Maria Strange, Tony Chater (New Island), the Robertson family (Port Stevens), Roddy and Lilly Napier, Michael and Jeanette Clarke (West Point Island), Rob and Lorraine McGill (Carcass Island), David and Suzan Pole-Evans (Saunders Island), Derek and Trudy Pettersson (Volunteer Beach), and many more kind and generous Falkland Islanders.

In Chile, Guillermo Luna-Jorquera and Alejandro Simeone helped us share the world of the Humboldt penguin.

We deeply appreciate the support of the Tristan da Cunha government, the South African Weather Service, Department of Environment and Tourism, South African National Antarctic Programme, and the Ovenstone Agencies for supporting our endeavour to reach the fabled islands of Tristan da Cunha and Gough, home of the amazing Northern rockhopper penguin. On and off the islands, we thank Julian Fitter, Henry Valentine, Dorrien Ven, Captain Tad de Oliveira, Captain Clarence October, Captain Peter Warren, Mike and Janice Hentley, Chris Bates, Jimmy and Felicity Glass, Eric McKenzie, Claire Volkwyn, and the team at the Gough Island Weather Station. We also wish to thank the tour companies who, between them over the years, opened doors to some truly magical penguin places: Lindblad Travel, Clipper Cruises, Quark Expeditions, Adventure Life, Galapagos Travel and especially Aurora Expeditions of Sydney and all the amazing staff of the *Polar Pioneer*.

In general, our deep appreciation goes to the dedicated conservation teams of many nations — whether NGOs or government-supported — labouring tirelessly to protect penguins and their environments and who actively facilitated our work. We are particularly grateful to the New Zealand Department of Conservation (especially for entrusting us with special one-of-a-kind permits that spanned the many months of our sailing ventures to the NZ Subantarctic Islands), the Galapagos National Park, the Charles Darwin Foundation (Galapagos), Phillip Island Nature Parks (Australia), Bruny Island (Tasmania), Pohatu Penguins (Banks Peninsula, NZ), Penguin Place (Dunedin, NZ), Little Blue Penguin Trust (NZ), Yellow-eyed Penguin Trust (NZ), the Southern African Foundation for the Conservation of Coastal Birds (SANCCOB), Tristan da Cunha Conservation Department, Falkland Conservation, and many, many more. All of these efforts are driven by visionary individuals who strive against innumerable odds to save penguins — your work is inspirational.

The wonderful Ellis family of Earth Sea Sky outdoor clothing, Jane and David with sons Ben, Mike and, tragically departed, Mark, have kept each of us warm and dry on many a penguin adventure with their superior expedition gear, made in New Zealand.

We are grateful to the Alexander Turnbull Library, Wellington, New Zealand and the Museum of New Zealand Te Papa Tongarewa, and in particular Alan Tennyson, Becky Masters and Jean-Claude Stahl, for producing the custom photo on page 159.

For your outstanding direct contributions to this book, we wish to thank: Daniel Ksepka, Yvon Le Maho, Matthew Shawkey, Sanne Boessenkool, David Ainley, Heather Lynch, Rory P. Wilson, Kyle W. Morrison, Barbara Wienecke, André Chiaradia, Hernán Vargas, Peter Ryan, P. Dee Boersma, David Thompson, Connie Glass, James Glass, Tina Glass, Ewan Fordyce, Julia Clarke, Sarah Crofts, Colin Monteath, Katrine Herrian, Dave Houston, Liliana D'Alba, Bastiaan Star, Laëtitia Kernaleguen, Sean Burns, Estelle van der Merwe, Tertius Gous, Ian Gaffney.

Last, but certainly not least, we thank Paul Bateman, our publisher, for espousing this project with faith and dedication, and, above all, Tracey Borgfeldt, associate publisher and editor extraordinaire, for your friendship, tenacity and unparalleled attention to detail.

We are indebted to many more people, but as this project spans several decades it is not possible to mention each and every person by name. Please know that even if you do not appear on this page, your contributions have not been forgotten.

INDEX